Geobotany Studies

Basics, Methods and Case Studies

The series includes outstanding monographs and collections of papers on a range of topics in the following fields: Phytogeography, Phytosociology, Plant Community Ecology, Biocoenology, Vegetation Science, Eco-informatics, Landscape Ecology, Vegetation Mapping, Plant Conservation Biology, and Plant Diversity. Contributions should reflect the latest theoretical and methodological developments or present new applications on large spatial or temporal scales that will reinforce our understanding of ecological processes acting at the phytocoenosis and vegetation landscape level. Case studies based on large data sets are also considered, provided they support habitat classification refinement, plant diversity conservation or vegetation change prediction. ***Geobotany Studies: Basics, Methods and Case Studies*** is the successor to *Braun-Blanquetia*, a journal published by the University of Camerino from 1984 to 2011 in cooperation with the *Station Internationale de Phytosociologie* (Bailleul, France) and the *Dipartimento di Botanica ed Ecologia* (Università di Camerino, Italy) and under the aegis of the *Société Amicale Francophone de Phytosociologie*, the *Société Française de Phytosociologie*, the *Rheinold-Tüxen-Gesellschaft* and the *Eastern Alpine and Dinaric Society for Vegetation Ecology*. This series promotes the expansion, evolution, and application of the invaluable scientific legacy of the Braun-Blanquet school.

More information about this series at http://www.springer.com/series/10526

Alessandro Alessandrini • Giampaolo Balboni •
Lisa Brancaleoni • Renato Gerdol •
Giovanni Nobili • Mauro Pellizzari •
Filippo Piccoli • Michele Ravaglioli

The Vascular Flora of the Bosco della Mesola Nature Reserve (Northern Italy)

 Springer

Alessandro Alessandrini
Istituto per i Beni Artistici
Culturali e Naturali RER
Bologna, Italy

Giampaolo Balboni
WWF Ferrara
Ferrara, Italy

Lisa Brancaleoni
Dipartimento di Scienze della
Vita e Biotecnologie
Università di Ferrara
Ferrara, Italy

Renato Gerdol
Dipartimento di Scienze della
Vita e Biotecnologie
Università di Ferrara
Ferrara, Italy

Giovanni Nobili
Carabinieri per la Biodiversità
di Punta Marina
Punta Marina (RA), Italy

Mauro Pellizzari
Istituto Comprensivo 'Bentivoglio'
Poggio Renatico (FE), Italy

Filippo Piccoli
Via Borgoleoni
Ferrara, Italy

Michele Ravaglioli
Carabinieri per la Biodiversità
di Punta Marina
Punta Marina (RA), Italy

ISSN 2198-2562   ISSN 2198-2570   (electronic)
Geobotany Studies
ISBN 978-3-030-63414-8   ISBN 978-3-030-63412-4   (eBook)
https://doi.org/10.1007/978-3-030-63412-4

This Springer imprint is published by the registered company Springer Nature Switzerland AG.
The registered company address is: Gewerbestrasse 11, 6330 Cham, Switzerland

Contents

1 Introduction 1
  1.1 Physical Environment and Climate 1
  1.2 Vegetation 4
  1.3 History and Management 17
  1.4 Materials and Methods 21

2 Flora 23
  2.1 History of the Botanical Exploration 23
  2.2 The Checklist 24
  2.3 Synthesis and Conclusion 72

Appendix A: List of the Ellenberg's indicator values.
X = undetermined 79

Appendix B: List of fungi recorded in the Bosco della Mesola Nature
Reserve (from Bernicchia et Corbetta 1982; Comune di Mesola 1988) 97

References 103

Chapter 1
Introduction

1.1 Physical Environment and Climate

The Bosco della Mesola (44°50′ N, 12°15′ E, 1058 ha), also called Boscone della Mesola or Gran Bosco della Mesola, is one of the main relics of lowland forest currently preserved in Italy. It is located in the southern deltaic lobe of Po, the longest Italian river, close to the Adriatic coast in North-Eastern Italy (Fig. 1.1). The Bosco della Mesola lies on geologically very recent terrain, consisting in fine-grained aeolian dune fields of late Medieval-Renaissance age (Stefani and Vincenzi 2005). Ground morphology is characterized by a system of topographic lows and highs formed of dune ridges with maximum elevation of about 7 m above sea level and dune slacks with minimum elevation of about 2 m below sea level. The orientation of the dune ridges is about 50° N, recording the direction of the dominant north-eastern winter wind, the so-called 'bora' (Fig. 1.1). Water supply to the nature reserve mainly originates from a network of canals providing freshwater from surrounding areas. However, the local aquifer is in hydraulic connection with the sea. This area has experienced strong subsidence especially since the half of the twentieth century when the subsidence rates strongly increased mainly due to anthropic causes, especially extraction of methane-bearing water, overexploitation of freshwater and riverbed deepening (Simeoni and Corbau 2009; Corbau et al. 2019). This implied a loss in altimetric elevation of about 2.5 m (Bondesan et al. 1995) and almost quintupled the rates of salt-wedge intrusion from the sea. Saltwater currently affects the topographic lows in the south-eastern part of the Nature Reserve (Gerdol et al. 2018).

The soils are overall poorly evolved with A-C profile, moderately alkaline pH, quite high carbonate content and rather poor nutrient contents. Most of the Bosco della Mesola area is characterized by entisols (Soil Survey Staff 1960): xeropsamments on high dune ridges, mesic aquic ustipsamments on flat sandy areas and typic psammaquents in dune slacks (IDROSER 1985; https://geo.regione.emilia-romagna.it/cartpedo). Inceptisoils are found in the low areas affected

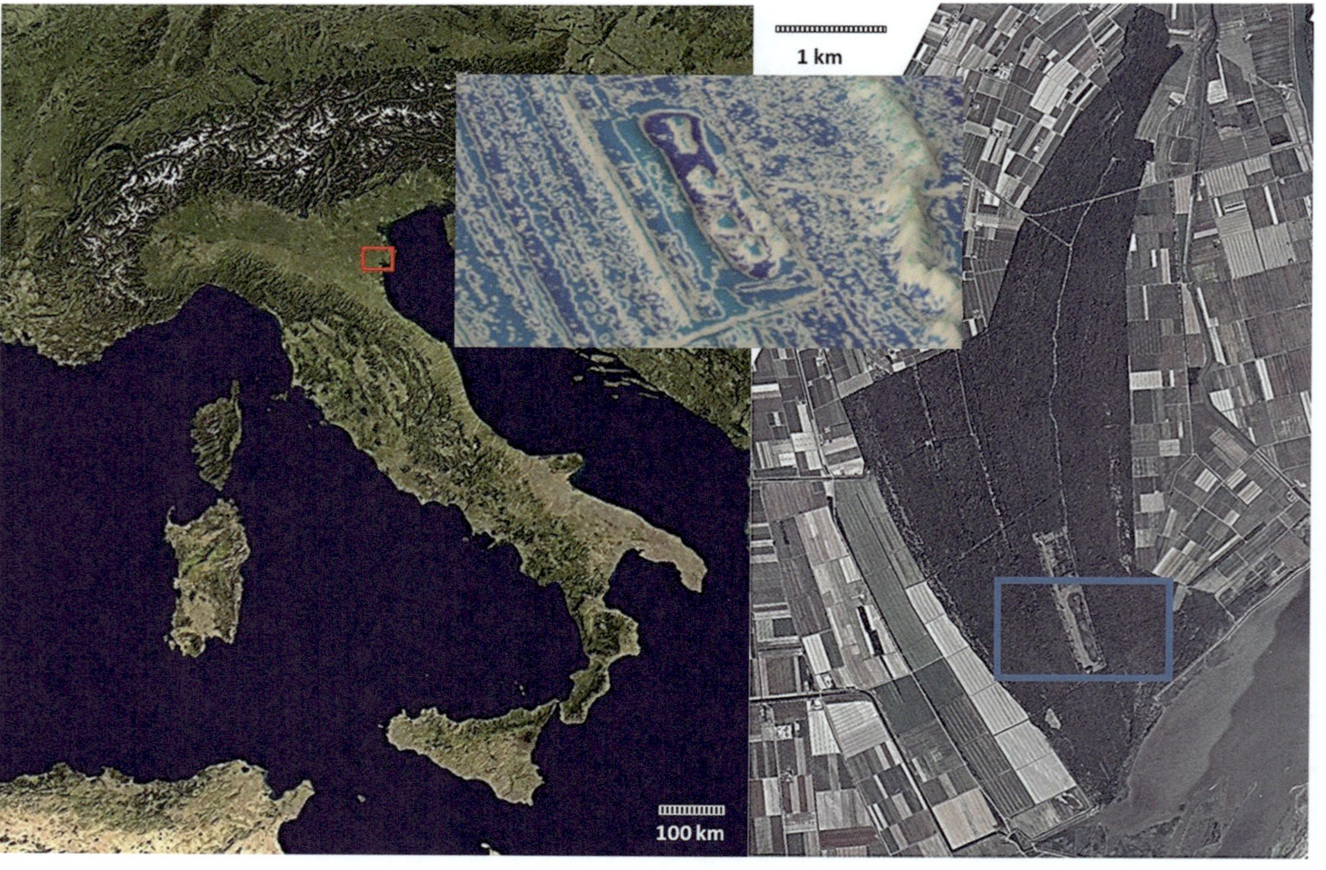

Fig. 1.1 Geographic location of the Bosco della Mesola Nature Reserve in Italy (red rectangle). Aerial picture of the Nature Reserve (courtesy of Ferrara Province) with 3D view of the area (courtesy of C. Corbau) in the blue rectangle

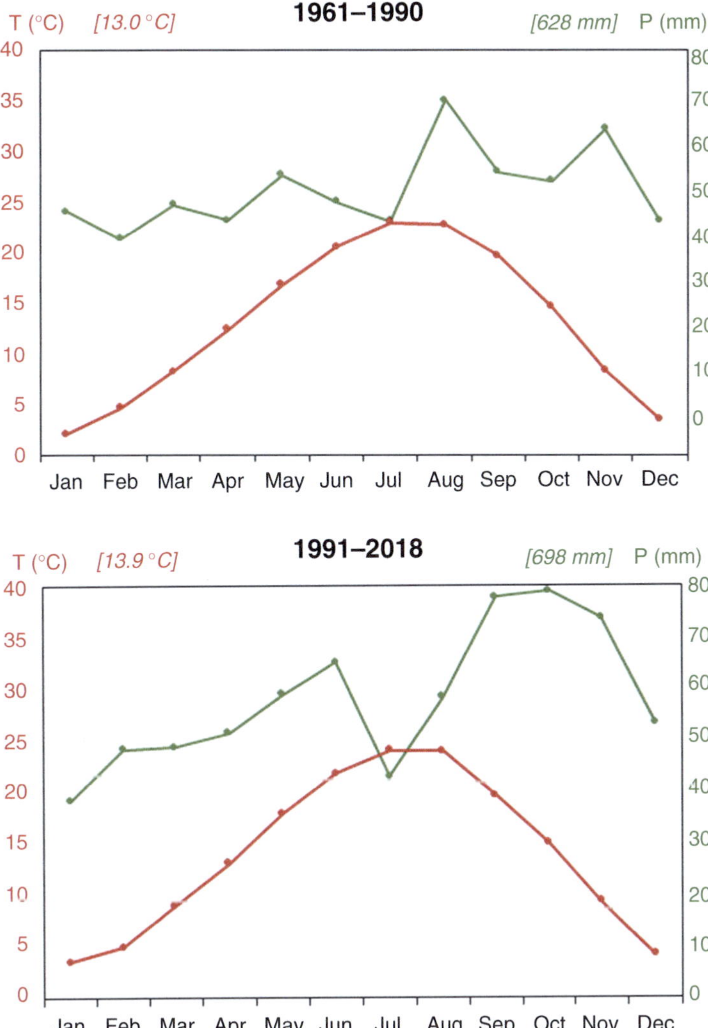

Fig. 1.2 Bagnouls-Gaussen ombrothermic diagrams for the periods 1961–1990 (upper panel) and 1991–2018 (lower panel)

by salt-wedge intrusion. These soils (halaquepts) possess high NaCl content and high electrical conductivity throughout the profile.

According to the Köppen classification of climates, the climate at Bosco della Mesola is temperate continental with mean annual temperature of 13–14 °C and temperature of the coldest month (January) between −1 and 3 °C. Mean total annual precipitation is 600–700 mm, rather homogeneously spread across seasons (Fig. 1.2). Based on the Rivas-Martinez phytoclimatic indices (http://www.globalbioclimatics.org; Table 1.1) the Bosco della Mesola Nature Reserve belongs

Table 1.1 Rivas-Martinez phytoclimatic indices for the periods 1961–1990 and 1991–2018 calculated using climatic data from the Mesola weather station (https://arpaeprv.datamb.it/dataset/erg5-eraclito)

	1961–1990	1991–2018
Annual ombrothermic index (Iov)	2.54	2.10
Continentality index (Ic)	21	20.5
Thermicity index (It)	380	414
Annual ombrothermic index (Io)	4.02	4.20

to the temperate climatic region with Iov > 2. The climate is semicontinental (Ic = *ca.* 21), lower thermotemperate (It = *ca.* 400), lower subhumid (Io = *ca.* 4). The mean annual temperature increased by about 1 °C from the 1961–1990 period to the present days (Fig. 1.2). Total annual precipitation slightly increased but monthly precipitation in July decreased in recent years. In association with higher monthly temperature, this results in a weak period of summer aridity that did not occur in the previous 30 years (Fig. 1.2).

1.2 Vegetation

The Bosco della Mesola Nature Reserve is mainly covered (*ca.* 90%) by woodland vegetation. Mantle and fringe vegetation is very rare and, therefore, not treated. Grassland vegetation (dry grasslands and wet meadows) covers small areas in canopy gaps. Hygrophilous vegetation (reeds and brackish wetlands) is relatively frequent, especially in dune slacks. Waterplant vegetation is widespread in aquatic habitats (canals and ponds). Ruderal vegetation is relatively frequent in disturbed areas such as track edges and areas trampled by wildlife.

Woodlands

QUERCETEA ILICIS Br.-Bl. in Br.-Bl., Roussine et Nègre 1952
Quercetalia ilicis Br.-Bl. ex Molinier 1934
Fraxino orni—Quercion ilicis Biondi, Casavecchia et Gigante in Biondi, Allegrezza, Casavecchia, Galdenzi, Gigante et Pesaresi 2013
Vincetoxico hirundinariae—Quercetum ilicis Gamper, Filesi, Buffa et Sburlino 2008

Holm-oak (*Quercus ilex*) woodlands (Fig. 1.3), settled on the highest dunes of recent age in the peripheral sectors of Bosco della Mesola, are included in the *Vincetoxico hirundinariae—Quercetum ilicis*, an association occurring in small fragmented areas along the North Adriatic coast, from the Tagliamento river mouth to Romagna. This association has notable phytogeographic value because of its extra-zonal location.

Fig. 1.3 In the foreground: *Cladio marisci—Fraxinetum oxycarpae* in a wet dune slack. In the background: *Vincetoxicum hirundinariae—Quercetum ilicis* on a high arid dune (Photo L. Brancaleoni)

Holm-oak is the dominant species in the tree layer with few other species, especially flowering ash (*Fraxinus ornus*). Most of the holm-oak trees are polycormic, because the forest was managed as coppice for long time. Tree renewal is strongly hindered by deer browsing. In the northern sector of Bosco della Mesola, due to the historical lack of fallow deer, a lower shrub layer is present with *Ruscus aculeatus*, *Asparagus acutifolius* and locally *Osyris alba*. The field layer is poor, with *Viola suavis*, *Teucrium capitatum*, *T. chamaedrys* and *Carex liparocarpos* as the most abundant species. From the dynamic point of view, the *Vincetoxico hirundinariae—Quercetum ilicis* is a stable association that tends to evolve in the long term towards mixed deciduous woodlands (Piccoli et al. 1983).

QUERCO-FAGETEA Br.-Bl. et Vlieger in Vlieger 1937
Quercetalia pubescenti-petraeae Klika 1933
Carpinion orientalis Horvat 1958 (= *Ostryo—Carpinion orientalis* Horvat 1959)
Quercus robur—Carpinus betulus community Piccoli, Gerdol et Ferrari 1983

Mixed deciduous woodlands cover the largest part of Bosco della Mesola on flat terrain originated by erosion of ancient dunes or filling of dune slacks (Piccoli and Gerdol 1980; Fig. 1.4). The tree layer is characterized by common oak (*Quercus robur*), oriental hornbeam (*Carpinus orientalis*), common hornbeam (*Carpinus betulus*). Holm-oak, field elm (*Ulmus minor*), narrow-leaf ash (*Fraxinus angustifolia* subsp. *oxycarpa*) and silver poplar (*Populus alba*) are also present in the tree layer, although with low abundance. The most frequent species in the shrub layer are *Acer campestre*, *Crataegus monogyna*, *Cornus mas*, *Ligustrum vulgare*, *Rhamnus cathartica* and *Frangula alnus*, all heavily damaged by deer browsing that also hinders renewal of these species.

The syntaxonomical typification of this woodland community is still unclear. Indeed, the *Quercus robur—Carpinus betulus* community of Bosco della Mesola differs considerably from the common oak forests in the Po Plain (*Polygonato multiflori—Quercetum roboris*; Bini et al. 2002) as well as from the Central-European *Carpinion betuli* woodland communities. On the other hand, the *Quercus robur—Carpinus betulus* community has closer similarity with amphiadriatic communities of the *Carpinion orientalis* (Blasi et al. 2004; Taffetani and Biondi 1995), as supported by the occurrence of several Southeastern-European—Pontic species, such as *Fraxinus angustifolia* subsp. *oxycarpa*, *Carpinus orientalis*, *Cornus mas*, *Rhamnus cathartica* and *Buglossoides purpurocaerulea*. The *Quercus robur—Carpinus betulus* community is the most mature woodland community at Bosco della Mesola and, consequently, dynamically stable.

ALNETEA GLUTINOSAE Br.-Bl. et R. Tüxen ex Westhoff, Dijk et Passchier 1946
Alnetalia glutinosae R. Tüxen 1937

Fig. 1.4 *Quercus robur—Carpinus betulus* community on a flat sandy area (Photo L. Brancaleoni)

Frangulo alni—Fraxinion oxycarpae Poldini, Sburlino et Venanzoni in Biondi, Allegrezza, Casavecchia, Galdenzi, Gasparri, Pesaresi, Poldini, Sburlino, Vagge et Venanzoni 2015

Cladio marisci—Fraxinetum oxycarpae Piccoli, Gerdol et Ferrari 1983 ex Piccoli 1995

Swamp woodlands, settled in dune slacks where fresh water table emerges seasonally (Fig. 1.3). Narrow-leaf ash tree woodlands are rather common in coastal dune systems along the North Adriatic coast for example at Punte Alberete and in Ravenna Pinewoods. It also occurs along the Tyrrhenian coast at the Circeo National Park (Piccoli and Gerdol 1980; Pedrotti and Gafta 1996; Stanisci et al. 1998; Merloni and Piccoli 2001).

The dominant species in the tree layer is narrow-leaf ash tree, often associated with field elm, white poplar and common oak. The shrub layer is composed of *Frangula alnus, Pyrus communis* subsp. *pyraster, Crataegus monogyna* and *Cornus sanguinea.* The field layer is characterized by hygrophilous species, especially *Cladium mariscus, Thelypteris palustris, Hydrocotyle vulgaris, Molinia arundinacea, Mentha aquatica* and *Euphorbia lucida,* an extremely rare species. The *Cladio marisci—Fraxinetum oxycarpae* suffers less damage from deer browsing compared to the other woodland communities. Dynamically, the *Cladio marisci— Fraxinetum oxycarpae* is unstable and closely related to high moisture level in the soil. Lowering of the water table, due to natural or anthropic causes, triggers dynamic processes that make this association evolve towards the *Quercus robur— Carpinus betulus* community (Piccoli et al. 1983).

RHAMNO—PRUNETEA Rivas Goday et Borja Carbonell ex R. Tüxen 1962
Prunetalia spinosae R. Tüxen 1952
Pruno—Rubion ulmifolii O. Bolós 1954
Viburno lantanae—Phillyreetum angustifoliae Gamper, Filesi, Buffa et Sburlino 2008

At Bosco della Mesola, only the holm-oak woodlands are bordered by a well developed shrubby edge, corresponding to the *Viburno lantanae—Phillyreetum angustifoliae.* This community is rather frequent on dune ridges along the North Adriatic coast. *Phillyrea angustifolia* and *Ruscus aculeatus* usually are the dominant species, generally associated with *Juniperus communis, Ligustrum vulgare, Osyris alba* and *Rubus ulmifolius.*

Grasslands

Dry grasslands, mostly formed of annual plants, occur in open flat sandy areas (Fig. 1.5). This vegetation belongs to the *Koelerio glaucae—Corynephoretea canescentis* Klika in Klika et Novák 1941 but can locally evolve towards the *Tortulo—Scabiosetum* Pignatti 1952 (*Artemisio—Koelerietalia albescentis* Sissingh

Fig. 1.5 Dry grassland (*Koelerio glaucae—Corynephoretea canescentis*) in open sandy areas (Photo L. Brancaleoni)

1974, *Syntrichio ruraliformis—Lomelosion argenteae* Biondi, Sburlino et Theurillat in Sburlino, Buffa, Filesi, Gamper et Ghirelli 2013). Occasionally, dry grasslands represent degradation stages closely related to the *Bromo tectorum—Phleetum arenarii* Korneck 1974 (*Corynephoretalia canescentis* Klika 1934, *Koelerion arenariae* R.Tx. 1937 corr. Gutermann et Mucina 1993).

Wetter open areas (Fig. 1.6) host an association mostly formed of perennial tussock-forming species (*Schoeno nigricantis—Erianthetum ravennae* Pignatti 1953: *Molinio—Arrhenateretea* R. Tüxen 1937, *Holoschoenetalia vulgaris* Br.-Bl. ex Tchou 1948, *Molinio—Holoschoenion vulgaris* Br.-Bl. ex Tchou 1948).

Wetlands

Reeds (*Phragmito—Magnocaricetea* Klika in Klika et Novák 1941) occur in two areas of Bosco della Mesola. The Elciola artificial pond (Fig. 1.7) hosts a mosaic of communities dominated by *Phragmites australis*, *Typha angustifolia*, *Schoenoplectus tabernaemontani* and *Juncus subnodulosus* (*Phragmitetalia communis* Koch 1926, *Phragmition communis* Koch 1926). The presence of *Juncus maritimus*, *J. littoralis* and *J. acutus* (Fig. 1.8) indicates some degree of saltwater influx (Gerdol et al. 2018). Open areas in the deepest dune slacks (Fig. 1.9) are usually covered by the *Mariscetum serrati* Zobrist 1935 (*Magnocaricetalia elatae* Pignatti 1953, *Magnocaricion elatae* Koch 1926). The dominant species, *Cladium mariscus*, is overall rare in lowland habitats in Northern Italy and indicates high levels of carbonate content in the soil and in the water.

Very rare and restricted to wet slacks and puddles is the *Cyperetum flavescentis* Koch ex Aich. 1933, an ephemeral community (*Isoeto—Nanojuncetea* Br.-Bl. et R. Tüxen ex Westhoff, Dijk, Passchier et Sissingh 1946, *Nanocyperetalia flavescentis* Klika 1935, *Nanocyperion flavescentis* Koch ex Libbert 1932).

Wetlands in the south-eastern part of the Bosco della Mesola Nature Reserve, experiencing saltwater influx due to intrusion of the salt wedge, are characterized by brackish vegetation types (Fig. 1.10). These mainly consist in rush communities (*Juncetea maritimi* Br.-Bl. in Br.-Bl., Roussine et Nègre 1952) and, secondarily, in annual glasswort beds (*Thero—Suaedetea splendentis* Rivas Martinez 1972) or cordgrass meadows (*Spartinetea glabrae* R. Tüxen in Beeftink 1962).

Aquatic Habitats

Waterplant vegetation is rather abundant in Bosco della Mesola, especially in canals (Gerdol et al. 2018). Two water plant communities freely floating on the surface of the water or within the waterbody belong to the class *Lemnetea minoris* Bolòs et Masclans 1955. Duckweed vegetation is formed of several autochthonous duckweed species (especially *Lemna minor* and *Spirodela polyrrhiza*), but recently the neo-tropical *Lemna minuta* became invasive in aquatic habitats (Fig. 1.11). This species tolerates shading and low temperature better than other duckweeds (Pellizzari and Piccoli 2001). *Ceratophyllum demersum* is the by far dominant species floating

Fig. 1.6 *Schoeno nigricantis—Erianthetum ravennae* in open wet areas (Photo L. Brancaleoni)

Fig. 1.7 Reed (*Phragmito—Magnocaricetea*) along the shore of the Elciola pond (Photo L. Brancaleoni)

under the water surface of the canals: the association is the *Ceratophylletum demersi* Hild 1956 (*Lemnetalia minoris* Bolòs et Masclans, *Lemno minoris—Hydrocharition morsus-ranae* Rivas-Martinez, Fernandez-Gonzalez et Loidi 1999).

Fig. 1.8 A reed stand rich in *Juncus acutus*, indicating saltwater influx (Photo L. Brancaleoni)

Rooting waterplant vegetation (class *Potametea pectinati* Klika in Klika et Novák 1941) is found in the deepest canals, with *Potamogeton* sp.pl. and *Myriophyllum spicatum* as the dominant species (Fig. 1.12). Only the Elciola pond hosts variable stands of *Utricularia australis* (*Utricularietum neglectae* T. Müller et Görs 1960: *Utricularietalia minoris* Den Hartog et Segal 1964, *Utricularion vulgaris* Passarge 1964). A community of stonewort algae (*Chara* sp.) is found in some watering pools fed with rainwater. *Ranunculus peltatus* subsp. *baudotii*, extremely rare all over this region, is recorded in this community.

Fig. 1.9 *Mariscetum serrati* in an open wet dune slack (Photo L. Brancaleoni)

Fig. 1.10 Brackish wetland with *Juncus littoralis* and *J. maritimus* (Photo M. Pellizzari)

Fig. 1.11 Freely floating waterplant vegetation (*Lemna minuta* community) in a canal (Photo M. Pellizzari)

Fig. 1.12 Rooting waterplant vegetation with *Potamogeton nodosus* as the dominant species (Photo L. Brancaleoni)

Ruderal Habitats

Two ruderal communities are found in areas subject to disturbance and/or heavy trampling (Fig. 1.13). These associations belong to the classes *Molinio— Arrhenateretea* R. Tüxen 1937 (*Cynodon dactylon* community of *Plantaginetalia majoris* R. Tüxen et Preising 1950), and *Artemisietea vulgaris* Lohmeyer, Preising et R. Tüxen ex von Rochow 1951 (*Dauco—Melilotion* Görs 1966), respectively.

1.3 History and Management

The Bosco della Mesola probably owes its preservation to being surrounded by waterbodies and lagoons for long time, so that it was of difficult access from the mainland. The Bosco della Mesola passed under several owners: from the House of Este since late Medieval times, to the House of Habsburg (1758), the Papal States (1785), the French Republic (1797), the Hospital S. Spirito in Sassia in Rome and, about a century ago (1919), to the Company for the reclamation of Ferrara lands. At

Fig. 1.13 Open area, heavy trampled by deer, covered by ruderal vegetation (*Cynodon dactylon* community) (Photo L. Brancaleoni)

that time the Bosco della Mesola essentially played a productive role to obtain firewood and, secondarily, timber. So, the wooded areas were prevalently managed as matriculated coppice. Non-wooded areas were grazed and, to a lesser extent, converted to vineyards. The Bosco della Mesola was the subject of intense hunting activity aimed at hunting deer, fallow deer, hare and several birds (especially pheasant and marsh birds). In the 1920s the Bosco della Mesola underwent the first activities directed to reforestation of degraded areas and afforestation of open areas, especially transplantation of pine on sand dunes.

In the early 1930s interventions of reforestation and forest management became more and more frequent. Furthermore, game density increased substantially after a period of hunting ban. In 1933, an unpublished technical report by Pavari (Zemello 1991) described the Bosco della Mesola as a scarcely matriculated holm-oak coppice with sparse undergrowth because of game browsing. Common oak grew only in cooler areas while narrow-leaf ash, black alder and white poplar, associated with swamp species in the understorey, were located in wetter areas. The lack of natural renewal from seed and the consequent absence of seedlings were mainly ascribed to the pressure exerted by large-size ungulates. Hence, it was suggested to reduce the ungulate load to about two individuals per ha. It was also suggested to thicken the woodlands by transplanting pine trees in the largest clearings and by managing the coppice with at least 50–60 matricines per ha in order to ensure adequate ground coverage. On the other hand, converting wooded areas into tall forest was discouraged because the concomitant effects of shallow water table and strong 'bora' winds can cause early drying of the crowns. At the outbreak of the Second World War, the management guidelines provided by experts were replaced by intense withdrawals of firewood 'as needed' by the local population, i.e. by uncontrolled cuts involving about 600 ha of woodlands during 1940–1945. In this period, more than a thousand adult pine trees were felled by order of the German occupation troops. In addition, hunting was allowed without restrictions which significantly decreased game density compared to the pre-war years.

The first forest management plan of the Bosco della Mesola was officially commissioned by the Company for the reclamation of Ferrara lands to Patrone (1948). In that period the demand for firewood was still strong. Accordingly, firewood production was recognized as the main function of the Bosco della Mesola. The coppice cutting shift was set at 25 years with matriculation of 50 plants per ha. This management plan also included plantation of holm oak in the clearings and plantation of hygrophilous species (black alder, greater ash and locally common oak) in more humid areas. In 1954, after the forest risked disappearing as a consequence of a land reform act which provided for expropriation and subsequent conversion into arable land, the ownership of the Bosco della Mesola passed to the Agency for State Forests. From that moment on, the management criteria changed radically as a result of the recognition of the great naturalistic value of this area. The new criteria were firstly mirrored in the settlement plan of Longhi (1967) who highlighted some drawbacks of Patrone's management guidelines. The main critical point stressed in the Longhi's management plant were: degradation of holm-oak woodlands and excessive spread of oriental hornbeam after coppicing, mainly caused by late frosts;

development of shrub species that compete with trees and thus slow down the dynamic evolution of the forest. The main objective set in that plan was the gradual conversion of the woodlands into a tall uneven forest without the need to reach a real 'normal state', but mainly focusing on renewal and conservation of the forest. The goal of the Longhi plan was to obtain natural renewal in woodland clearings, greater in wetter and cooler areas and smaller in xeric areas. In that period the entire perimeter of the Bosco della Mesola was fenced to avoid the escape of ungulates and reduce the entry of poachers and unauthorized visitors. Consequently, the relict native deer population, representing one of the outstanding naturalistic components of this area, gradually increased from about 40 individuals in 1970 to 250–300 individuals in 2019.

In 1972 an Integral Nature Reserve (Bassa dei Frassini—Balanzetta; 222 ha) was established in the central-western part of the Bosco della Mesola. No interventions of any kind are allowed in this area and access is permitted only for study and research purposes. In 1977 a State Nature Reserve was established over the entire area of the Bosco della Mesola (1058 ha). The subsequent forest management plan (Minerbi 1984), defined technical guidelines for implementing management tools suitable for conservation purposes. Minerbi observed that excessive density of trees hampers natural renewal in large part of the wooded areas. So, he set up a curve to identify optimal densities of trees of various species in order to guarantee sufficient renewal in the understorey. The novelty of Minerbi's plan consisted in a multi-disciplinary approach to forest management at Bosco della Mesola. The multi-disciplinary studies carried out at that time were compiled in a report ranging from hydrological, to pedological and zoological aspects besides forestry (IDROSER 1985). A few years later (1989) the Bosco della Mesola was included in the Po Delta Regional Park. The latest forest management plan (Naccarato 2004) was still based on a naturalistic multi-disciplinary approach, with particular attention to hydraulic man-agement and conservation of the threatened native deer population. A recent Life Natura 2000 project entitled 'Conservation of habitats and species of the SIC BOSCO DELLA MESOLA' (LIFE 00 NAT/17/7147) aimed to define guidelines for interventions directed to the conservation of habitats and species of priority importance in the Bosco della Mesola (Directive 92/43 UE). The main critical points outlined in this project were: worsening of the quality of surface waters and salinisation of the aquifer triggered by subsidence and land reclamation; excessive grazing and/or browsing by the exotic fallow deer which can exert negative effects on land tortoise (*Testudo hermanni*) and, especially, on the native deer population. Therefore, both the level and quality of ground and surface waters were monitored periodically from 2002 to 2004. This led to the creation of a mathematical model to describe the dynamics of water table and water flows in the Bosco della Mesola. Based on the model outcomes, hydraulic facilities were created for intercepting and reshaping the hydraulic flows. In addition, the canal network was subjected to extraordinary maintenance works by expanding and reshaping the canals section wherever appropriate. Ordinary maintenance consisted in cleaning canals that tended to be filled by organic and inorganic debris. Artificial pools were created to encour-age the colonization of amphibian populations and ephemeral plant species. To counteract and mitigate ungulate overload, the reserve was divided into wildlife management units by creating temporary fences. Preliminary observations revealed promising results in terms of holm-oak renewal. The latest forest management plan and the Life Natura 2000 project stimulated the activation of procedures for controlling the density of the fallow deer population. This resulted in increased plant cover in areas formerly grazed by fallow deer.

1.4   Materials and Methods

The checklist of the Bosco della Mesola vascular flora was compiled on the basis of many field surveys carried out from 1979 to 2019. The surveys were performed in all seasons over the whole range of habitats occurring in the Nature Reserve. Particular care was devoted to marginal and ruderal habitats, ofted neglected during botanical explorations. The specimens collected during these surveys were stored in the Herbarium of Ferrara University (FER). The data obtained by field surveys were integrated by thorough literature search. This eventually resulted in a list including *ca.* 3000 units. The data were arranged in a database (created with MS Access), where every unit contained the name of the taxon as quoted in the original source, the year of observation/collection or publication, besides any annotation useful for subsequent data elaboration. The original checklist was linked to an Excel sheet containing information about life form, chorotype, habitat preference and ecological requirement. The taxonomic nomenclature follows http://www.theplantlist.org/ (updates in http://www.worldfloraonline.org/) and Euro+Med PlantBase http://www.emplantbase.org/ for *Anisantha*, *Brachypodium* and *Molinia*. The nomenclature of old records was updated according to the current source. Only occasionally could nomenclatural updating not be performed. Life forms were in accordance with Ellenberg and Mueller-Dombois (1965/1966), with the exception of helophytes (Tiner 1999). Chorotypes (see Fattorini (2017) for conceptual framework and definition) were in accordance with Pignatti (2005) for native species and Galasso et al. (2018) for alien species. Alien species were split into naturalized, casual or invasive. Occurrence of 'feral' alien species was specified as well (see Galasso et al. 2018 for terminology). Additional information was provided for taxa of peculiar interest, at the regional or national scale. Habitat preference was defined according to the main types described in Sect. 1.2: Woodlands, Dry grasslands, Wet meadows, Reeds, Brackish wetlands, Aquatic habitats and Ruderal habitats. Several species occur across more than one habitat type. In those cases, we opted to restrict the reference to habitat preference to the two habitat types in which the species occurs most frequently. Ecological requirements were defined using Ellenberg's indicator values drawn from Pignatti (2005) for native species and Domina et al. (2018) for alien species.

The text version of the checklist contained only part of the information included in the table version. The species were grouped into families, with all taxa listed in alphabetic order. Synonyms were given only in case of potential terminological confusion. The taxa recorded during our surveys were considered as currently existing and highlighted in bold character. The symbol ★ indicates taxa not cited in previous records and, therefore, new to the Bosco della Mesola flora. Conversely, the taxa recorded in literature only and not found during our field surveys are in normal character. Literature records for these taxa were chronologically separated in two groups: taxa recorded in the Piccoli's (1964) checklist (P) and taxa recorded in previous, more ancient, literature references (A). Information was given for life form, chorotype, habitat preference as well as for Ellenberg's indicator values (see Appendix A).

The floristic data compiled in the checklist were elaborated with the main objective to detect temporal changes in the Bosco della Mesola flora. The comparison was made between the previous checklist, dating back to the 1960s (Piccoli 1964; for brevity referred to as '1960') and the checklist of taxa recorded in our surveys (referred to as '2020'). To this aim, the frequencies of taxa grouped according to life form, chorotype, habitat preference and ecological requirements were calculated for the two checklists. Life forms and chorotypes were grouped in synthetic categories as follows.

Life forms Chamaephytes, Geophytes, Helophytes, Hemicryptophytes, Hydrophytes, Lianas, Phanerophytes and Therophytes.

Chorotypes Atlantic, Boreal, Cosmopolitan, Endemic, Eurasiatic, Mediterranean, SE European—W Asiatic, Temperate, Tropical and Aliens.

We finally compiled data for conservation status of the protected species.

Supporting environmental data consisted in daily minimum and maximum temperature and daily precipitatation for the period 1961–2018 at the closest weather station (Mesola; https://arpaeprv.datamb.it/dataset/erg5-eraclito). This data set was obtained by high-resolution interpolation of climatic data collected at a number of weather station across Emilia-Romagna and neighbouring regions (Antolini et al. 2016). We used these data to calculate climatic indices separately for the periods 1961–1990 and 1991–2018. This allowed to compare floristic variations with temporal climatic changes.

Chapter 2
Flora

2.1 History of the Botanical Exploration

The botanical exploration of Bosco della Mesola started at the beginning of the Twentieth Century. The first stages of the exploration were part of pioneer floristic research over the whole Ferrara Province (Revedin 1909) and the coastal territory (Béguinot 1910, 1916). These two authors listed many species and provided general information about Bosco della Mesola and surrounding areas. More general overviews about flora, vegetation, environment and forestry in Bosco della Mesola were published about 40 years later (Patrone 1951; Carullo 1953; Albanese 1958; Jedlowski 1960).

In the early 1960s, after a thorough botanical exploration, Piccoli (1964) compiled the first checklist of vascular plants for the Bosco della Mesola. Thereafter, the Bosco della Mesola was the subject of studies addressing several aspects of the flora and vegetation of this area (Agostini 1965; Stampi 1966a, b; Longhi 1968; Zangheri 1971; Corbetta 1973; Minerbi et al. 1975; Corbetta and Pettenei 1976; Debolini and Ricceri 1976; Cencini 1979). Typology, ecology and dynamics of forest vegetation in the Bosco della Mesola was studied in detail during the 1980s (Piccoli and Gerdol 1980, 1990; Gerdol et al. 1985; Piccoli 1987). In those times the first vegetation map of the Bosco della Mesola was published (Piccoli et al. 1983).

In the last three decades several studies addressed different aspects of the flora and vegetation of the Bosco della Mesola: forest typology and ecology (Ferrari 1994; Buffa et al. 1994; Taffetani and Biondi 1995; Gamper et al. 2008); vegetation mapping (Zemello 1991; Piccoli 1995; Piccoli et al. 1996, 1999); flora and vegetation of aquatic habitats (Pellizzari and Piccoli 2001; Piccoli and Pellizzari 2003); relationships between vegetation and herpetofauna (Mazzotti et al. 2007). The most recent botanical studies in Bosco della Mesola addressed: taxonomy of the genus *Viola* (Pellizzari and Brancaleoni 2016); ecology of wetland vegetation (Gerdol et al. 2018); experimental translocation of the endangered species *Kosteletzkya pentacarpos* (Brancaleoni et al. 2018).

A. Alessandrini et al., *The Vascular Flora of the Bosco della Mesola Nature Reserve (Northern Italy)*, Geobotany Studies, https://doi.org/10.1007/978-3-030-63412-4_2

The mycological flora of the Bosco della Mesola is quite well known thanks to investigations carried out mostly in the 1980s (Bernicchia and Corbetta 1982; Bernicchia 1983, 1986; Ginns and Bernicchia 1984; Bernicchia et al. 1987; Bernicchia and Intini 1988). A list of fungal species recorded in the Bosco della Mesola Nature Reserve (Bernicchia and Corbetta 1982; Comune di Mesola 1988) is compiled in Appendix B. Besides the forest management plans (see Chap. 1) several aspects of forest management were addressed by IDROSER (1985), Minerbi (1985), Caramalli (1987a, b), Pellizzari and Pagnoni (1998), Mazzotti et al. (2003a, b) and Naccarato (2004). Scientific knowledge of Bosco della Mesola, with special attention to botanical aspects, was disseminated to the public in a number of non-technical articles (Cori and Raminelli 1982; Bonani 1989; Raminelli et al. 1993; Rossi 1994; Pagnoni 1998, 1999; Cortesi 2005; Piccoli 2005).

2.2   The Checklist

Alismataceae

Alisma lanceolatum With.—Helophyte, Subcosmopolitan; Reeds.
Alisma plantago-aquatica L.—Helophyte, Subcosmopolitan; Reeds.

Amaranthaceae

(A, P) *Amaranthus albus* L.—Scapose Therophyte, Naturalized Alien; Ruderal habitats.
Amaranthus deflexus L.—Scapose Therophyte, Invasive Alien; Ruderal habitats.
Amaranthus retroflexus L.—Scapose Therophyte, Invasive Alien; Ruderal habitats.
Arthrocnemum macrostachyum (Moric.) Piirainen et G. Kadereit (*A. glaucum* (Moq.) Ung.-Sternb.)—Frutescent Chamaephyte (Succulent), Stenomediterranean; Brackish wetlands.
Atriplex portulacoides (L.) Aellen—Frutescent Chamaephyte, Boreal; Brackish wetlands.
Atriplex prostrata Boucher ex DC.—Scapose Therophyte, Boreal; Ruderal habitats.
(A, P) *Atriplex rosea* L.—Scapose Therophyte, Central Asiatic-Eurimediterranean; Ruderal habitats.
(A) *Atriplex tatarica* L.—Scapose Therophyte, South-European—South-Siberian; Ruderal habitats.
Bassia crassifolia (Pall.) Soldano [*Spirobassia hirsuta* (L.) Freitag et G. Kadereit]—Scapose Therophyte, South-European—South-Siberian; Brackish wetlands.
Bassia laniflora (S.G. Gmel.) A.J. Scott—Scapose Therophyte, South-European—South-Siberian; Dry grasslands.
Chenopodium album L.—Scapose Therophyte, Subcosmopolitan; Ruderal habitats. *C. album* belongs to a complex and little known group of species. We include herein records of *C. viride* that is sometimes considered as a different species.

(A, P) *Cycloloma atriplicifolium* (Spreng.) J.M. Coult.—Scapose Therophyte, Naturalized Alien; Dry grasslands.

Dysphania ambrosioides (L.) Mosyakin et Clemants (*Chenopodium a.* L.)—Scapose Therophyte, Naturalized Alien; Ruderal habitats.

Salicornia patula Willd.—Scapose Therophyte, Atlantic; Brackish wetlands.

Salsola soda L.—Scapose Therophyte, Palaeotemperate; Brackish wetlands.

(A, P) *Salsola tragus* L.—Scapose Therophyte, Palaeotemperate; Brackish wetlands.

Sarcocornia fruticosa (L.) A.J. Scott [*Arthrocnemum fruticosum* (L.) Moq.]—Frutescent Chamaephyte (Succulent), Eurimediterranean; Brackish wetlands.

Suaeda maritima (L.) Dumort.—Scapose Therophyte, Cosmopolitan; Brackish wetlands.

(A) *Suaeda vera* Forssk. ex J.F. Gmel.—Nanophanerophyte, Cosmopolitan; Brackish wetlands.

Amaryllidaceae

Allium vineale L.—Bulbous Geophyte, Eurimediterranean; Ruderal habitats.

Apiaceae

(A, P) *Anthriscus caucalis* M. Bieb.—Scapose Therophyte, Palaeotemperate; Ruderal habitats.

★ ***Apium graveolens*** L.—Scapose Hemicryptophyte, Palaeotemperate; Wet meadows.

Berula erecta (Huds.) Coville—Rhizome Geophyte, Boreal; Reeds.

(A, P) *Bupleurum tenuissimum* L.—Scapose Therophyte, Eurimediterranean; Brackish wetlands.

Daucus carota L.—Scapose Hemicryptophyte, Palaeotemperate; Dry grasslands, Ruderal habitats.

(A, P) *Echinophora spinosa* L.—Scapose Hemicryptophyte, Eurimediterranean; Dry grasslands.

(A, P) *Eryngium maritimum* L.—Rhizome Geophyte, Eurimediterranean-Atlantic; Dry grasslands.

★ ***Oenanthe lachenalii*** C.C. Gmel.—Scapose Hemicryptophyte, Eurimediterranean-Subatlantic; Reeds.

(A, P) *Oenanthe pimpinelloides* L.—Scapose Hemicryptophyte, Eurimediterranean-Subatlantic; Wet meadows.

Peucedanum palustre (L.) Moench [*Thysselinum palustre* (L.) Hoffm.]—Scapose Hemicryptophyte, Euro-Siberian; Wet meadows.

Torilis nodosa (L.) Gaertn.—Scapose Therophyte, Eurimediterranean-Turanian; Ruderal habitats.

Apocynaceae

(A) *Apocynum venetum* L.—Rhizome Geophyte, South-European—South-Siberian; Woodlands. The *A. venetum* populations in the region Emilia-Romagna represent the southernmost outpost of the distribution range of this species in Italy.

Fig. 2.1 *Vincetoxicum hirundinaria* subsp. *hirundinaria* (Photo A. Alessandrini)

Vincetoxicum hirundinaria Medik. subsp. ***hirundinaria***—Scapose Hemicryptophyte, Eurasiatic; Woodlands. The Bosco della Mesola population has a climbing habit. Based on the whole set of characters it, however, belongs to the nominal subspecies (Fig. 2.1).

Araceae

Lemna gibba L.—Errant Hydrophyte, Subcosmopolitan; Aquatic habitats.
Lemna minor L.—Errant Hydrophyte, Subcosmopolitan; Aquatic habitats.
Lemna minuta Kunth—Errant Hydrophyte, Invasive Alien; Aquatic habitats. In aquatic habitats this species is rapidly replacing the native duckweed species (Fig. 2.2).
Lemna trisulca L.—Errant Hydrophyte, Cosmopolitan; Aquatic habitats.
Spirodela polyrhiza (L.) Schleid.—Errant Hydrophyte, Subcosmopolitan; Aquatic habitats.

Araliaceae

Hedera helix L.—Liana, Eurimediterranean; Woodlands.
Hydrocotyle vulgaris L.—Rhizome Geophyte, European-Caucasian; Wet meadows. This species is becoming more and more rare. It currently occurs only in a few coastal stations in Emilia-Romagna and Veneto (Argenti et al. 2019) (Fig. 2.3).

Aristolochiaceae

Aristolochia clematitis L.—Root-budding Geophyte, Eurimediterranean; Wet meadows.

Fig. 2.2 *Lemna minuta* (Photo M. Pellizzari)

Fig. 2.3 *Hydrocotyle vulgaris* (Photo A. Alessandrini)

Fig. 2.4 *Ruscus aculeatus* (Photo A. Alessandrini)

Aristolochia rotunda L.—Bulbous Geophyte, Eurimediterranean; Woodlands, Wet meadows.

Asparagaceae

Asparagus acutifolius L.—Nanophanerophyte, Stenomediterranean; Woodlands.
Asparagus officinalis L.—Rhizome Geophyte, Eurimediterranean; Naturalized Alien (Feral); Dry grasslands, Ruderal habitats.
Ruscus aculeatus L.—Frutescent Chamaephyte, Eurimediterranean; Woodlands (Fig. 2.4).

Asteraceae

Fig. 2.5 *Centaurea calcitrapa* (Photo R. Gerdol)

Achillea collina (Becker ex Rchb. f.) Heimerl—Scapose Hemicryptophyte, Southeast-European; Dry grasslands. The occurrence of *A. millefolium* has not been ascertained at Bosco della Mesola.

★ *Achillea nobilis* L.—Scapose Hemicryptophyte, South-European—South-Siberian; Dry grasslands.

★ *Achillea roseoalba* Ehrend.—Scapose Hemicryptophyte, Central-European; Ruderal habitats.

(A) *Ambrosia maritima* L.—Scapose Therophyte, Eurimediterranean; Dry grasslands. Extremely rare species, probably extinct all over Italy (Bartolucci et al. 2018).

Ambrosia psilostachya DC.—Rhizome Geophyte, Invasive Alien; Dry grasslands.

(A) *Artemisia alba* Turra—Suffrutescent Chamaephyte, North-Eurimediterranean; Dry grasslands.

Artemisia caerulescens L. subsp. *caerulescens*—Suffrutescent Chamaephyte, Eurimediterranean; Brackish wetlands.

(A, P) *Artemisia campestris* L.—Suffrutescent Chamaephyte, Boreal; Dry grasslands.

Artemisia vulgaris L.—Scapose Hemicryptophyte, Boreal; Ruderal habitats.

Bellis perennis L.—Scapose Hemicryptophyte (Rosette), European-Caucasian; Ruderal habitats.

(A, P) *Bidens tripartitus* L. subsp. *tripartitus*—Scapose Therophyte, Eurasiatic; Wet meadows, Ruderal habitats.

Carduus acanthoides L.—Scapose Hemicryptophyte, European-Caucasian; Ruderal habitats.

(A, P) *Carduus nutans* L. subsp. *nutans*—Scapose Hemicryptophyte, Atlantic; Ruderal habitats.

★ *Carduus pycnocephalus* L. subsp. *pycnocephalus*—Scapose Hemicryptophyte, Eurimediterranean-Turanian; Ruderal habitats.

Centaurea calcitrapa L.—Scapose Hemicryptophyte, Eurimediterranean; Dry grasslands (Fig. 2.5).

Centaurea nigrescens Willd.—Scapose Hemicryptophyte, Eurasiatic; Ruderal habitats.

Centaurea tommasinii A. Kern.—Scapose Hemicryptophyte, Endemic; Dry grasslands.

Chondrilla juncea L.—Scapose Hemicryptophyte, South-European—South-Siberian; Dry grasslands, Ruderal habitats.

Cichorium intybus L.—Scapose Hemicryptophyte, Palaeotemperate; Ruderal habitats.

Cirsium arvense (L.) Scop.—Root-budding Geophyte, Eurasiatic; Ruderal habitats.

Cirsium vulgare (Savi) Ten.—Scapose Hemicryptophyte, Palaeotemperate; Ruderal habitats.

(A) *Crepis foetida* L.—Scapose Therophyte, Eurimediterranean; Ruderal habitats.

Crepis neglecta L.—Scapose Therophyte, Northeast-Eurimediterranean; Ruderal habitats.

(A) *Crepis pulchra* L.—Scapose Therophyte, Eurimediterranean; Ruderal habitats.

★ ***Crepis sancta*** (L.) Babc. subsp. ***nemausensis*** (P. Fourn.) Babc.—Scapose Therophyte, Naturalized Alien; Ruderal habitats.

Crepis setosa Haller f.—Scapose Therophyte, East-Eurimediterranean; Ruderal habitats.

(A) *Crepis vesicaria* L. subsp. *taraxacifolia* (Thuill.) Thell.—Scapose Therophyte, Eurimediterranean-Subatlantic; Ruderal habitats.

Dittrichia viscosa (L.) Greuter subsp. ***viscosa***—Scapose Hemicryptophyte, Eurimediterranean; Ruderal habitats.

Erigeron canadensis L.—Scapose Therophyte, Invasive Alien; Ruderal habitats.

Eupatorium cannabinum L.—Scapose Hemicryptophyte, Palaeotemperate; Ruderal habitats, Reeds.

Galactites tomentosus Moench—Scapose Hemicryptophyte, Stenomediterranean; Dry grasslands.

(A, P) *Helichrysum italicum* (Roth) G. Don—Suffrutescent Chamaephyte, North-Eurimediterranean; Dry grasslands.

Helminthotheca echioides (L.) Holub—Scapose Therophyte, Eurimediterranean; Ruderal habitats.

Hypochaeris glabra L.—Scapose Therophyte, Eurimediterranean; Dry grasslands.

Hypochaeris radicata L.—Scapose Hemicryptophyte (Rosette), European-Caucasian; Dry grasslands, Ruderal habitats (Fig. 2.6).

(A, P) *Inula salicina* L.—Scapose Hemicryptophyte, European-Caucasian; Wet meadows.

(A) *Jacobaea erratica* (Bertol.) Fourr. (*Senecio e.* Bertol.; *Jacobaea aquatica* sensu Auct. ital.)—Scapose Hemicryptophyte, Central-European; Wet meadows.

(A, P) *Lactuca saligna* L.—Scapose Therophyte, Eurimediterranean-Turanian; Ruderal habitats.

Lactuca serriola L.—Scapose Hemicryptophyte, South-European—South-Siberian; Ruderal habitats.

(A, P) *Laphangium luteoalbum* (L.) Tzvelev—Scapose Therophyte, Subcosmopolitan; Wet meadows.

(A, P) *Lapsana communis* L. subsp. *communis*—Scapose Therophyte, Palaeotemperate; Ruderal habitats.

Fig. 2.6 *Hypochaeris radicata* (Photo R. Gerdol)

Leontodon hispidus L.—Scapose Hemicryptophyte (Rosette), European-Caucasian; Dry grasslands, Ruderal habitats.

Leontodon saxatilis Lam. subsp. ***saxatilis***—Scapose Therophyte, Eurimediterranean; Wet meadows.

Limbarda crithmoides (L.) Dumort.—Scapose Hemicryptophyte, European-Caucasian; Brackish wetlands.

Matricaria chamomilla L.—Scapose Therophyte, Subcosmopolitan; Ruderal habitats.

Onopordum acanthium L.—Scapose Hemicryptophyte, East-Mediterranean-Montane; Ruderal habitats.

Picris hieracioides L.—Scapose Hemicryptophyte, Euro-Siberian; Ruderal habitats.

(A, P) *Pilosella officinarum* Vaill.—Scapose Hemicryptophyte (Rosette), European-Caucasian; Dry grasslands.

(A, P) *Pilosella piloselloides* (Vill.) Soják—Scapose Hemicryptophyte, European-Caucasian; Dry grasslands.

Pulicaria dysenterica (L.) Bernh.—Scapose Hemicryptophyte, Eurimediterranean; Ruderal habitats; Reeds.

(A, P) *Pulicaria vulgaris* Gaertn.—Scapose Therophyte, Palaeotemperate; Wet meadows.

Scolymus hispanicus L. subsp. ***hispanicus***—Scapose Hemicryptophyte, Eurimediterranean; Ruderal habitats (Fig. 2.7).

Senecio inaequidens DC.—Scapose Therophyte, Naturalized Alien; Ruderal habitats.

Senecio vulgaris L.—Scapose Therophyte, Eurimediterranean; Ruderal habitats.

★ ***Silybum marianum*** (L.) Gaertn.—Scapose Hemicryptophyte, Eurimediterranean-Turanian; Ruderal habitats.

(A, P) *Solidago canadensis* L.—Scapose Hemicryptophyte, Naturalized Alien; Wet meadows, Reeds.

Solidago gigantea Aiton—Scapose Hemicryptophyte, Naturalized Alien; Wet meadows, Reeds.

Fig. 2.7 *Scolymus hispanicus* (Photo R. Gerdol)

Sonchus asper (L.) Hill—Scapose Therophyte, Eurasiatic; Ruderal habitats.

Sonchus maritimus L.—Scapose Hemicryptophyte (Rosette), Eurimediterranean; Brackish wetlands.

Sonchus oleraceus L.—Scapose Therophyte, Eurasiatic; Ruderal habitats.

★ *Symphyotrichum squamatum* (Spreng.) G.L. Nesom—Scapose Therophyte, Naturalized Alien; Ruderal habitats.

(A, P) *Taraxacum fulvum* (gr.)—Scapose Hemicryptophyte (Rosette), Palaeotemperate; Dry grasslands.

Taraxacum officinale (gr.)—Scapose Hemicryptophyte (Rosette), Boreal; Ruderal habitats.

(A, P) *Taraxacum palustre* (gr.)—Scapose Hemicryptophyte (Rosette), Eurasiatic; Wet meadows.

(A, P) *Tragopogon orientalis* L.—Scapose Hemicryptophyte, Euro-Siberian; Dry grasslands.

★ *Tragopogon porrifolius* L.—Scapose Hemicryptophyte, Eurimediterranean; Dry grasslands, Ruderal habitats.

(A, P) *Tragopogon pratensis* L.—Scapose Hemicryptophyte, Euro-Siberian; Dry grasslands, Ruderal habitats.

Tripolium pannonicum (Jacq.) Dobrocz.—Scapose Hemicryptophyte, Eurasiatic; Brackish wetlands (Fig. 2.8).

(A, P) *Tussilago farfara* L.—Rhizome Geophyte, Palaeotemperate; Wet meadows.

Xanthium orientale L. subsp. *italicum* (Moretti) Greuter—Scapose Therophyte, Invasive Alien; Dry grasslands.

Xanthium spinosum L.—Scapose Therophyte, Naturalized Alien; Ruderal habitats.

(A, P) *Xanthium strumarium* L.—Scapose Therophyte, Naturalized Alien; Dry grasslands.

Berberidaceae

Fig. 2.8 *Tripolium pannonicum* (Photo R. Gerdol)

Fig. 2.9 *Carpinus orientalis* subsp. *orientalis* (Photo A. Alessandrini)

Berberis vulgaris L.—Nanophanerophyte, Eurasiatic; Woodlands.

Betulaceae

Alnus glutinosa (L.) Gaertn.—Phanerophyte, Palaeotemperate; Woodlands.
Carpinus betulus L.—Phanerophyte, European-Caucasian; Woodlands.
Carpinus orientalis Mill. subsp. ***orientalis***—Phanerophyte, Pontic; Woodlands (Fig. 2.9).

Boraginaceae

(A, P) *Anchusa azurea* Mill.—Scapose Hemicryptophyte, Eurimediterranean; Ruderal habitats.

(A) *Anchusa officinalis* L.—Scapose Hemicryptophyte, Pontic; Dry grasslands.

Buglossoides purpurocaerulea (L.) I.M. Johnst.—Scapose Hemicryptophyte, Pontic; Woodlands.

(A, P) *Cynoglossum creticum* Mill.—Scapose Hemicryptophyte, Eurimediterranean; Dry grasslands.

Cynoglossum officinale L.—Scapose Hemicryptophyte, Eurasiatic; Dry grasslands.

Echium vulgare L.—Scapose Hemicryptophyte, European; Ruderal habitats.

Heliotropium europaeum L.—Scapose Therophyte, Eurimediterranean; Ruderal habitats.

(A, P) *Lithospermum officinale* L.—Scapose Hemicryptophyte, Euro-Siberian; Ruderal habitats.

Myosotis arvensis (L.) Hill—Scapose Therophyte, European-Caucasian; Ruderal habitats.

(A, P) *Myosotis ramosissima* Rochel ex Schult.—Scapose Therophyte, European-Caucasian; Dry grasslands, Ruderal habitats.

Symphytum officinale L.—Scapose Hemicryptophyte, European-Caucasian; Wet meadows.

Brassicaceae

Alyssum alyssoides (L.) L.—Scapose Therophyte, Eurimediterranean; Dry grasslands.

Arabidopsis thaliana (L.) Heynh.—Scapose Therophyte, Palaeotemperate; Ruderal habitats.

Arabis hirsuta (L.) Scop.—Scapose Hemicryptophyte, European; Dry grasslands.

(A, P) *Arabis sagittata* (Bertol.) DC.—Scapose Hemicryptophyte, Southeast-European; Woodlands.

(A, P) *Arabis turrita* L. [*Pseudoturritis turrita* (L.) Al-Shehbaz]—Scapose Hemicryptophyte, South-European—South-Siberian; Woodlands.

(A, P) *Cakile maritima* Scop. subsp. *maritima*—Scapose Therophyte, Eurimediterranean-Subatlantic; Dry grasslands, Brackish wetlands.

Capsella bursa-pastoris (L.) Medik.—Scapose Hemicryptophyte, Cosmopolitan; Ruderal habitats.

★ *Capsella rubella* Reut.—Scapose Therophyte, Eurimediterranean; Ruderal habitats.

Cardamine hirsuta L.—Scapose Therophyte, Cosmopolitan; Ruderal habitats.

Cardamine pratensis L.—Scapose Hemicryptophyte, European; Wet meadows.

★ *Cardaria draba* (L.) Desv.—Rhizome Geophyte, Eurimediterranean-Turanian; Ruderal habitats.

★ ***Descurainia sophia*** (L.) Webb ex Prantl—Scapose Therophyte, Palaeotemperate; Ruderal habitats.

Diplotaxis tenuifolia (L.) DC.—Scapose Hemicryptophyte, Subatlantic; Ruderal habitats.

Erophila verna (L.) DC.—Scapose Therophyte, Boreal; Ruderal habitats.

Fig. 2.10 *Lomelosia argentea* (Photo A. Alessandrini)

(A) *Eruca vesicaria* (L.) Cav.—Scapose Therophyte, Eurimediterranean-Turanian; Ruderal habitats.

Hornungia petraea (L.) Rchb.—Scapose Therophyte, Eurimediterranean; Dry grasslands.

(A, P) *Raphanus raphanistrum* L.—Scapose Therophyte, Eurimediterranean; Ruderal habitats.

★ ***Rapistrum rugosum*** (L.) Arcang.—Scapose Therophyte, Eurimediterranean; Ruderal habitats.

Sisymbrium officinale (L.) Scop.—Scapose Therophyte, Palaeotemperate; Ruderal habitats.

Cannabaceae

Humulus lupulus L.—Liana, European-Caucasian; Woodlands, Ruderal habitats.

Caprifoliaceae

★ ***Dipsacus fullonum*** L.—Scapose Hemicryptophyte, Eurimediterranean; Ruderal habitats.

(A, P) *Knautia arvensis* (L.) Coult.—Scapose Hemicryptophyte, Eurasiatic; Dry grasslands.

Lomelosia argentea (L.) Greuter et Burdet—Scapose Hemicryptophyte, South-European—South-Siberian; Dry grasslands (Fig. 2.10).

Lonicera caprifolium L.—Liana, South-European—South-Siberian; Woodlands.

Lonicera etrusca Santi—Liana, Eurimediterranean; Woodlands.

(A, P) *Scabiosa triandra* L.—Scapose Therophyte, South-European—South-Siberian; Ruderal habitats.

(A) *Valeriana officinalis* L.—Scapose Hemicryptophyte, European; Wet meadows.

Valerianella locusta (L.) Laterr.—Scapose Therophyte, Eurimediterranean; Ruderal habitats.

Caryophyllaceae

(A) *Agrostemma githago* L.—Scapose Therophyte, European-Caucasian; Ruderal habitats.

Arenaria leptoclados (Rchb.) Guss.—Scapose Therophyte, Palaeotemperate; Ruderal habitats.

Arenaria serpyllifolia L. subsp. *serpyllifolia*—Scapose Therophyte, Subcosmopolitan; Ruderal habitats.

Cerastium glomeratum Thuill.—Scapose Therophyte, Eurimediterranean; Ruderal habitats.

(A, P) *Cerastium holosteoides* Fr.—Scapose Hemicryptophyte, Eurasiatic; Ruderal habitats.

(A, P) *Cerastium ligusticum* Viv.—Scapose Therophyte, West-Stenomediterranean; Ruderal habitats.

Cerastium semidecandrum L.—Scapose Therophyte, Eurasiatic; Ruderal habitats.

Minuartia hybrida (Vill.) Shischk. subsp. *hybrida* [*Sabulina tenuifolia* (L.) Rchb. subsp. *t.*]—Scapose Therophyte, Palaeotemperate; Ruderal habitats.

★ *Minuartia mediterranea* (Ledeb. ex Link) K. Malý [*Sabulina m.* (Ledeb. ex Link) Rchb.]—Scapose Therophyte, Northwest-Mediterranean-Montane; Dry grasslands, Ruderal habitats.

Petrorhagia saxifraga (L.) Link—Caespitose Hemicryptophyte, Eurimediterranean; Dry grasslands.

★ *Polycarpon tetraphyllum* (L.) L. subsp. *diphyllum* (Cav.) O. Bolòs et Font Quer—Scapose Therophyte, Eurimediterranean; Ruderal habitats.

★ *Polycarpon tetraphyllum* (L.) L. subsp. *tetraphyllum*—Scapose Therophyte, Eurimediterranean; Ruderal habitats.

Saponaria officinalis L.—Scapose Hemicryptophyte, Euro-Siberian; Dry grasslands, Ruderal habitats.

(A, P) *Silene colorata* Poir.—Scapose Therophyte, Stenomediterranean; Dry grasslands.

Silene conica L.—Scapose Therophyte, Palaeotemperate; Dry grasslands.

(A, P) *Silene dioica* (L.) Clairv.—Scapose Hemicryptophyte, Palaeotemperate; Woodlands.

Silene italica (L.) Pers.—Scapose Hemicryptophyte (Rosette), Eurimediterranean; Woodlands.

Silene latifolia Poir. subsp. *alba* (Mill.) Greuter et Poiret—Scapose Hemicryptophyte, Stenomediterranean; Ruderal habitats.

Silene vulgaris (Moench) Garcke—Scapose Hemicryptophyte, Palaeotemperate; Ruderal habitats.

(A, P) *Spergularia media* (L.) C. Presl—Suffrutescent Chamaephyte, Subcosmopolitan; Brackish wetlands.

Stellaria media (L.) Vill. subsp. *media*—Reptant Therophyte, Cosmopolitan; Ruderal habitats.

Fig. 2.11 *Ceratophyllum demersum* (Photo A. Alessandrini)

★ ***Stellaria pallida*** (Dumort.) Crép.—Scapose Therophyte, Palaeotemperate; Ruderal habitats.

Celastraceae

Euonymus europaeus L.—Phanerophyte, Eurasiatic; Woodlands.

Ceratophyllaceae

Ceratophyllum demersum L.—Rooted Hydrophyte, Subcosmopolitan; Aquatic habitats (Fig. 2.11).

Cistaceae

Cistus creticus L. subsp. ***eriocephalus*** (Viv.) Greuter et Burdet— Nanophanerophyte, West-Stenomediterranean; Dry grasslands (Fig. 2.12).

(A, P) *Cistus salviifolius* L.—Nanophanerophyte, Stenomediterranean; Dry grasslands.

(A, P) *Fumana procumbens* (Dunal) Gren. et Godr.—Suffrutescent Chamaephyte, Eurimediterranean-Pontic; Dry grasslands.

★ ***Helianthemum apenninum*** (L.) Mill.—Suffrutescent Chamaephyte, Atlantic; Dry grasslands (Fig. 2.13).

(A, P) *Helianthemum jonium* Lacaita—Suffrutescent Chamaephyte, Endemic; Dry grasslands.

Helianthemum nummularium (L.) Mill.—Suffrutescent Chamaephyte, European-Caucasian; Dry grasslands.

Commelinaceae

★ ***Commelina communis*** L.—Bulbous Geophyte, Naturalized Alien (Feral); Ruderal habitats.

Fig. 2.12 *Cistus creticus* subsp. *eriocephalus* (Photo L. Brancaleoni)

Fig. 2.13 *Helianthemum apenninum* (Photo R. Gerdol)

Convolvulaceae

Calystegia sepium (L.) R. Br.—Liana, Palaeotemperate; Ruderal habitats.

(A, P) *Calystegia soldanella* (L.) Roem. et Schult.—Rhizome Geophyte, Cosmopolitan; Dry grasslands.

Convolvulus arvensis L.—Rhizome Geophyte, Palaeotemperate; Ruderal habitats.

(A) *Cuscuta campestris* Yunck.—Reptant Therophyte (Parasite), Invasive Alien; Dry grasslands.

(A, P) *Cuscuta epithymum* (L.) L.—Reptant Therophyte (Parasite), Eurasiatic; Dry grasslands.

★ ***Dichondra micrantha*** Urb.—Rhizome Geophyte, Naturalized Alien (Feral); Ruderal habitats.

Cornaceae

Cornus mas L.—Phanerophyte, South-European—South-Siberian; Woodlands (Fig. 2.14).

Fig. 2.14 *Cornus mas* (Photo R. Gerdol)

Cornus sanguinea L. subsp. *hungarica* (Kárpáti) Soó—Phanerophyte, Eurasiatic; Woodlands.

Crassulaceae

Sedum acre L.—Frutescent Chamaephyte (Succulent), European-Caucasian; Dry grasslands, Ruderal habitats.
★ *Sedum album* L.—Frutescent Chamaephyte (Succulent), Eurimediterranean; Ruderal habitats.

Cupressaceae

Juniperus communis L.—Phanerophyte, Boreal; Dry grasslands.

Cyperaceae

Bolboschoenus maritimus (L.) Palla—Rhizome Geophyte, Cosmopolitan; Wet meadows, Reeds.
Carex acutiformis Ehrh.—Rhizome Geophyte, Eurasiatic; Wet meadows.
(A) *Carex caryophyllea* Latourr.—Scapose Hemicryptophyte, Eurasiatic; Woodlands, Dry grasslands.
Carex distans L.—Caespitose Hemicryptophyte, Eurimediterranean; Wet meadows.
Carex divisa Huds.—Rhizome Geophyte, Eurimediterranean-Atlantic; Wet meadows.
★ *Carex divulsa* Stokes—Caespitose Hemicryptophyte, Eurimediterranean; Wet meadows.
Carex extensa Gooden.—Caespitose Hemicryptophyte, Eurimediterranean-Subatlantic; Brackish wetlands.
Carex flacca Schreb.—Rhizome Geophyte, European; Woodlands.
Carex hirta L.—Rhizome Geophyte, European-Caucasian; Wet meadows.
Carex liparocarpos Gaudin—Rhizome Geophyte, Southeast-European; Woodlands, Dry grasslands.

Fig. 2.15 *Cladium mariscus* (Photo L. Brancaleoni)

Carex otrubae Podp.—Caespitose Hemicryptophyte, Eurimediterranean-Atlantic; Wet meadows.

Carex panicea L.—Rhizome Geophyte, Euro-Siberian; Wet meadows.

(A, P) *Carex pendula* Huds.—Caespitose Hemicryptophyte, Eurasiatic; Woodlands, Wet meadows.

★ *Carex punctata* Gaudin—Caespitose Hemicryptophyte, Eurimediterranean-Subatlantic; Wet meadows. Extremely rare in coastal areas.

★ *Carex riparia* Curtis—Rhizome Geophyte, Eurasiatic; Reeds.

Carex rostrata Stokes—Rhizome Geophyte, Boreal; Wet meadows.

Carex viridula Michx.—Caespitose Hemicryptophyte, Eurasiatic; Wet meadows.

Cladium mariscus (L.) Pohl—Rhizome Geophyte, Subcosmopolitan; Wet meadows, Brackish wetlands (Fig. 2.15).

(A, P) *Cyperus capitatus* Vand.—Rhizome Geophyte, Stenomediterranean; Dry grasslands.

Cyperus eragrostis Lam.—Scapose Therophyte, Naturalized Alien; Ruderal habitats.

Cyperus flavescens L.—Caespitose Therophyte, Subcosmopolitan; Wet meadows.

Cyperus fuscus L.—Caespitose Therophyte, Palaeotemperate; Wet meadows.

(A, P) *Cyperus glomeratus* L.—Helophyte, Palaeosubtropical; Wet meadows.

(A) *Cyperus laevigatus* L. subsp. *distachyos* (All.) Ball.—Rhizome Geophyte, Subcosmopolitan; Wet meadows.

Cyperus longus L.—Helophyte, Palaeotemperate; Wet meadows.

Eleocharis palustris (L.) Roem. et Schult.—Rhizome Geophyte, Subcosmopolitan; Wet meadows.

Schoenoplectus lacustris (L.) Palla—Helophyte, Subcosmopolitan; Reeds.

Schoenoplectus litoralis (Schrad.) Palla—Helophyte, Palaeosubtropical; Brackish wetlands.

Schoenoplectus mucronatus (L.) Palla—Helophyte, Pantropical; Wet meadows.

Schoenoplectus pungens (Vahl) Palla—Rhizome Geophyte, Subcosmopolitan; Brackish wetlands.

Schoenoplectus tabernaemontani (C.C. Gmel.) Palla—Helophyte, Euro-Siberian; Reeds.

(A) *Schoenoplectus triqueter* (L.) Palla—Helophyte, Boreal; Reeds.

Schoenus nigricans L.—Caespitose Hemicryptophyte, Subcosmopolitan; Wet meadows (Fig. 2.16).

Scirpoides holoschoenus (L.) Soják—Rhizome Geophyte, Stenomediterranean; Wet meadows.

Dennstaedtiaceae

Pteridium aquilinum (L.) Kuhn—Rhizome Geophyte, Cosmopolitan; Woodlands, Dry grasslands.

Dioscoreaceae

Dioscorea communis (L.) Caddick et Wilkin (*Tamus c.* L.)—Root-budding Geophyte, Eurimediterranean; Woodlands.

Elaeagnaceae

★ *Elaeagnus angustifolia* L.—Phanerophyte, Naturalized Alien (Feral); Ruderal habitats.

(A, P) *Hippophae fluviatilis* (Soest) Rivas Mart.—Phanerophyte, Eurasiatic; Dry grasslands.

Equisetaceae

Equisetum arvense L.—Rhizome Geophyte, Boreal; Ruderal habitats.

(A, P) *Equisetum fluviatile* L.—Rhizome Geophyte, Boreal; Wet meadows.

Equisetum palustre L.—Rhizome Geophyte, Boreal; Wet meadows.

Equisetum ramosissimum Desf.—Rhizome Geophyte, Boreal; Ruderal habitats.

Equisetum telmateia Ehrh.—Rhizome Geophyte, Boreal; Ruderal habitats.

Euphorbiaceae

Euphorbia cyparissias L.—Scapose Hemicryptophyte, Central-European; Dry grasslands.

Euphorbia esula L.—Scapose Hemicryptophyte, Euro-Siberian; Wet meadows.

Euphorbia helioscopia L. subsp. *helioscopia*—Scapose Therophyte, Cosmopolitan; Ruderal habitats.

Euphorbia lucida Waldst. et Kit.—Scapose Hemicryptophyte, South-European—South-Siberian; Wet meadows. This very rare species (Fig. 2.17) has been recorded for the first time in Italy at Bosco della Mesola (Debolini and Ricceri 1976). Subsequently, it was also recorded in Veneto (Busnardo 2000; Masin and Scortegagna 2012).

Euphorbia maculata L.—Scapose Therophyte, Naturalized Alien; Ruderal habitats.

Euphorbia palustris L.—Rhizome Geophyte, Euro-Siberian; Wet meadows, Reeds.

Fig. 2.16 *Schoenus nigricans* (Photo R. Gerdol)

Fig. 2.17 *Euphorbia lucida* (Photo L. Brancaleoni)

(A, P) *Euphorbia paralias* L.—Frutescent Chamaephyte, Eurimediterranean; Dry grasslands.

(A, P) *Euphorbia peplis* L.—Reptant Therophyte, Eurimediterranean; Dry grasslands.

Euphorbia peplus L.—Scapose Therophyte, Euro-Siberian; Dry grasslands.

Euphorbia platyphyllos L.—Scapose Therophyte, Eurimediterranean; Dry grasslands.

(A) *Euphorbia prostrata* Aiton—Reptant Therophyte, Invasive Alien; Ruderal habitats.

★ ***Mercurialis annua*** L.—Scapose Therophyte, Palaeotemperate; Ruderal habitats.

Fabaceae

★ ***Amorpha fruticosa*** L.—Phanerophyte, Invasive Alien (Feral); Ruderal habitats.

(A, P) *Astragalus glycyphyllos* L.—Reptant Chamaephyte, South-European—South-Siberian; Woodlands.

Colutea arborescens L.—Phanerophyte, Eurimediterranean; Woodlands.

Dorycnium herbaceum Vill.—Scapose Hemicryptophyte, South-European—South-Siberian; Dry grasslands.

Dorycnium hirsutum (L.) Ser.—Suffrutescent Chamaephyte, Eurimediterranean; Dry grasslands.

(A, P) *Dorycnium pentaphyllum* Scop.—Scapose Hemicryptophyte, South-European—South-Siberian; Dry grasslands.

Genista tinctoria L.—Suffrutescent Chamaephyte, Eurasiatic; Dry grasslands.

Hippocrepis emerus (L.) Lassen (*Emerus major* Mill., *Coronilla emerus* L.)—Nanophanerophyte, Central-European; Woodlands.

(A, P) *Lathyrus hirsutus* L.—Scapose Therophyte, Eurimediterranean; Ruderal habitats.

(A, P) *Lathyrus pratensis* L.—Scapose Hemicryptophyte, Palaeotemperate; Dry grasslands, Ruderal habitats.

★ ***Lathyrus sphaericus*** Retz.—Scapose Therophyte, Eurimediterranean; Dry grasslands, Ruderal habitats.

(A, P) *Lotus angustissimus* L.—Scapose Therophyte, Eurimediterranean; Dry grasslands.

Lotus corniculatus L.—Scapose Hemicryptophyte, Palaeotemperate; Ruderal habitats.

Lotus maritimus L. [*L. tetragonolobus* L., *Tetragonolobus m.* (L.) Roth]—Scapose Hemicryptophyte, Eurimediterranean; Wet meadows.

Lotus tenuis Waldst. et Kit. ex Willd.—Scapose Hemicryptophyte, Palaeotemperate; Wet meadows.

Medicago lupulina L.—Scapose Therophyte, Palaeotemperate; Ruderal habitats.

Medicago marina L.—Reptant Chamaephyte, Eurimediterranean; Dry grasslands.

Medicago minima (L.) L.—Scapose Therophyte, Eurimediterranean; Dry grasslands.

Medicago sativa L. subsp. *sativa*—Scapose Hemicryptophyte, Eurasiatic; Ruderal habitats.

Melilotus albus Medik.—Scapose Therophyte, Eurasiatic; Ruderal habitats.

Melilotus indicus (L.) All.—Scapose Therophyte, Eurimediterranean-Turanian; Ruderal habitats.

(A, P) *Melilotus neapolitanus* Ten.—Scapose Therophyte, Stenomediterranean; Dry grasslands, Ruderal habitats.

(A, P) *Melilotus officinalis* (L.) Pall.—Scapose Hemicryptophyte, Eurasiatic; Ruderal habitats.

Ononis natrix L. subsp. *natrix*—Caespitose Hemicryptophyte, Eurimediterranean; Dry grasslands.

Ononis spinosa L.—Suffrutescent Chamaephyte, Eurimediterranean; Dry grasslands.

Robinia pseudoacacia L.—Phanerophyte, Invasive Alien (Feral); Woodlands, Ruderal habitats.

(A, P) *Trifolium arvense* L.—Scapose Therophyte, Palaeotemperate; Dry grasslands.

Trifolium campestre Schreb.—Scapose Therophyte, Palaeotemperate; Dry grasslands.

Trifolium fragiferum L. subsp. *fragiferum*—Reptant Chamaephyte, Palaeotemperate; Ruderal habitats.

Trifolium nigrescens Viv.—Scapose Therophyte, Eurimediterranean; Dry grasslands, Ruderal habitats.

Trifolium pratense L.—Scapose Hemicryptophyte, Euro-Siberian; Dry grasslands, Ruderal habitats.

Fig. 2.18 *Quercus ilex* (Photo R. Gerdol)

Trifolium repens L.—Reptant Chamaephyte, Palaeotemperate; Dry grasslands, Ruderal habitats.
Trifolium scabrum L. subsp. ***scabrum***—Reptant Therophyte, Eurimediterranean; Dry grasslands.
Trifolium striatum L.—Scapose Therophyte, Palaeotemperate; Dry grasslands.
(A, P) *Vicia cracca* L.—Scapose Hemicryptophyte, Eurasiatic; Ruderal habitats.
Vicia sativa L.—Scapose Therophyte, Eurimediterranean-Turanian; Dry grasslands, Ruderal habitats.
Vicia villosa Roth subsp. ***ambigua*** (Guss.) Kerguelen (*Vicia pseudocracca* Bertol.)—Scapose Therophyte, Stenomediterranean; Dry grasslands.
Vicia villosa Roth subsp. ***varia*** (Host.) Corb.—Scapose Therophyte, Eurimediterranean; Dry grasslands, Ruderal habitats.

Fagaceae

Quercus ilex L.—Phanerophyte, Stenomediterranean; Woodlands (Fig. 2.18).
Quercus pubescens Willd. subsp. ***pubescens***—Phanerophyte, Southeast-European; Woodlands.
Quercus robur L. subsp. ***robur***—Phanerophyte, European-Caucasian; Woodlands.

Gentianaceae

Fig. 2.19 *Schenkia spicata* (Photo A. Alessandrini)

Blackstonia perfoliata (L.) Huds.—Scapose Therophyte, Eurimediterranean; Wet meadows.

Centaurium erythraea Rafn—Scapose Hemicryptophyte, Palaeotemperate; Wet meadows.

Centaurium pulchellum (Sw.) Druce—Scapose Therophyte, Palaeotemperate; Wet meadows.

Centaurium tenuiflorum (Hoffmanns. et Link) Fritsch—Scapose Therophyte, Palaeotemperate; Wet meadows, Brackish wetlands.

Schenkia spicata (L.) G. Mans.—Scapose Therophyte, Eurimediterranean; Wet meadows (Fig. 2.19).

Geraniaceae

Erodium cicutarium (L.) L'Hér.—Scapose Therophyte, Subcosmopolitan; Ruderal habitats.

★ ***Geranium columbinum*** L.—Scapose Therophyte, South-European—South-Siberian; Dry grasslands, Ruderal habitats.

★ *Geranium dissectum* L.—Scapose Therophyte, Eurasiatic; Ruderal habitats.

Geranium molle L.—Scapose Therophyte, Eurasiatic; Ruderal habitats.

★ *Geranium purpureum* Vill.—Scapose Therophyte, Eurimediterranean; Ruderal habitats.

Geranium pusillum L.—Scapose Therophyte, Eurasiatic; Ruderal habitats.

(A, P) *Geranium rotundifolium* L.—Scapose Therophyte, Palaeotemperate; Ruderal habitats.

Haloragaceae

Myriophyllum spicatum L.—Rooted Hydrophyte, Subcosmopolitan; Aquatic habitats.

Hydrocharitaceae

Najas minor All.—Rooted Hydrophyte, Palaeotemperate-Subtropical; Aquatic habitats.

Hypericaceae

Hypericum perforatum L. subsp. *perforatum*—Scapose Hemicryptophyte, Palaeotemperate; Ruderal habitats.

Iridaceae

Iris pseudacorus L.—Rhizome Geophyte, Eurasiatic; Wet meadows, Reeds.

Juglandaceae

★ *Juglans regia* L.—Phanerophyte, Casual Alien (Feral); Ruderal habitats.

Juncaceae

Juncus acutus L.—Caespitose Hemicryptophyte, Eurimediterranean; Brackish wetlands (Fig. 2.20).

Juncus articulatus L.—Rhizome Geophyte, Boreal; Reeds.

Juncus bufonius L.—Caespitose Therophyte, Cosmopolitan; Wet meadows.

Juncus compressus Jacq.—Rhizome Geophyte, Eurasiatic; Wet meadows, Brackish wetlands.

Juncus gerardii Loisel.—Rhizome Geophyte, Boreal; Wet meadows, Brackish wetlands.

★ *Juncus inflexus* L.—Caespitose Hemicryptophyte, Palaeotemperate; Wet meadows.

Juncus littoralis C.A. Mey.—Caespitose Hemicryptophyte, Eurimediterranean-Turanian; Brackish wetlands.

Juncus maritimus Lam.—Rhizome Geophyte, Subcosmopolitan; Brackish wetlands.

Juncus subnodulosus Schrank—Rhizome Geophyte, European-Caucasian; Reeds (Fig. 2.21).

★ *Juncus tenageja* Ehrh.—Caespitose Therophyte, Palaeotemperate; Wet meadows.

Fig. 2.20 *Juncus acutus* (Photo L. Brancaleoni)

Fig. 2.21 *Juncus subnodulosus* (Photo A. Alessandrini)

Luzula campestris (L.) DC.—Caespitose Hemicryptophyte, European-Caucasian; Woodlands, Dry grasslands.

Juncaginaceae

(A, P) *Triglochin maritimum* L.—Scapose Hemicryptophyte, Subcosmopolitan; Brackish wetlands.

Lamiaceae

(A, P) *Ajuga chamaepitys* (L.) Schreb. subsp. *chamaepitys*—Scapose Therophyte, Eurimediterranean; Ruderal habitats.
Ajuga reptans L.—Reptant Chamaephyte, European-Caucasian; Wet meadows, Ruderal habitats.

Ballota nigra L. subsp. ***foetida*** (Vis.) Hayek—Scapose Hemicryptophyte, Eurimediterranean; Ruderal habitats.

Clinopodium acinos (L.) Kuntze—Scapose Therophyte, Eurimediterranean; Dry grasslands.

Clinopodium nepeta (L.) Kuntze—Scapose Hemicryptophyte, South-European Orophyte; Dry grasslands.

Clinopodium vulgare L.—Scapose Hemicryptophyte, Boreal; Woodlands.

(A, P) *Galeopsis ladanum* L.—Scapose Therophyte, Eurasiatic; Ruderal habitats.

Glechoma hederacea L.—Reptant Chamaephyte, Boreal; Woodlands, Ruderal habitats.

Lamium amplexicaule L.—Scapose Therophyte, Palaeotemperate; Ruderal habitats.

Lamium purpureum L.—Scapose Therophyte, Eurasiatic; Ruderal habitats.

Lycopus europaeus L.—Scapose Hemicryptophyte, Palaeotemperate; Wet meadows.

(A, P) *Marrubium vulgare* L.—Scapose Hemicryptophyte, South-European—South-Siberian; Dry grasslands.

Mentha aquatica L.—Scapose Hemicryptophyte, Palaeotemperate; Wet meadows, Reeds.

(A) *Mentha suaveolens* Ehrh.—Scapose Hemicryptophyte, Eurimediterranean; Wet meadows.

(A, P) *Nepeta cataria* L.—Scapose Hemicryptophyte, East-Mediterranean-Turanian; Ruderal habitats.

(A, P) *Origanum vulgare* L.—Scapose Hemicryptophyte, Eurasiatic; Dry grasslands.

(A, P) *Prunella laciniata* (L.) L.—Scapose Hemicryptophyte, Eurimediterranean; Dry grasslands.

Prunella vulgaris L. subsp. ***vulgaris***—Scapose Hemicryptophyte, Boreal; Dry grasslands.

Salvia pratensis L.—Scapose Hemicryptophyte, Eurimediterranean; Dry grasslands, Ruderal habitats.

Salvia verbenaca L.—Scapose Hemicryptophyte, Stenomediterranean-Atlantic; Dry grasslands, Ruderal habitats.

★ ***Stachys palustris*** L.—Scapose Hemicryptophyte, Boreal; Wet meadows, Reeds.

(A, P) *Stachys recta* L.—Scapose Hemicryptophyte, North-Mediterranean-Montane; Dry grasslands.

Teucrium capitatum L.—Suffrutescent Chamaephyte, Stenomediterranean; Dry grasslands (Fig. 2.22).

Teucrium chamaedrys L.—Suffrutescent Chamaephyte, Eurimediterranean; Woodlands.

Teucrium scordium L. subsp. ***scordium***—Scapose Hemicryptophyte, European-Caucasian; Wet meadows.

Lauraceae

Fig. 2.22 *Teucrium capitatum* (Photo M. Pellizzari)

Laurus nobilis L.—Phanerophyte, Casual Alien (Feral); Woodlands, Ruderal habitats.

Lentibulariaceae

Utricularia australis R. Br.—Errant Hydrophyte, European; Aquatic habitats.

Linaceae

★ ***Linum bienne*** Mill.—Scapose Hemicryptophyte, Eurimediterranean; Dry grasslands.
(A, P) *Linum catharticum* L.—Scapose Therophyte, Eurimediterranean; Dry grasslands.
Linum maritimum L.—Scapose Hemicryptophyte, West-Stenomediterranean; Brackish wetlands.

Lythraceae

★ ***Lythrum hyssopifolia*** L.—Scapose Therophyte, Subcosmopolitan; Wet meadows.
Lythrum salicaria L.—Scapose Hemicryptophyte, Subcosmopolitan; Wet meadows, Ruderal habitats.

Malvaceae

Fig. 2.23 *Kosteletzkya pentacarpos* (Photo L. Brancaleoni)

Abutilon theophrasti Medik.—Scapose Therophyte, Naturalized Alien; Ruderal habitats.

Althaea officinalis L.—Scapose Hemicryptophyte, Southeast-European; Wet meadows.

Kosteletzkya pentacarpos (L.) Ledeb. (*Hibiscus pentacarpos* L.)—Scapose Hemicryptophyte, South-European—South-Siberian; Wet meadows, Reeds (Fig. 2.23). This species, listed in the EU Habitats Directive, is extremely rare in Italy where it currently occurs only at Bosco della Mesola and a few stations in nearby coastal areas in Veneto (Argenti et al. 2019).

Malva sylvestris L.—Scapose Hemicryptophyte, Euro-Siberian; Ruderal habitats.

Moraceae

★ ***Ficus carica*** L.—Phanerophyte, Eurimediterranean-Turanian; Ruderal habitats.

Morus alba L.—Phanerophyte, Naturalized Alien (Feral); Ruderal habitats.

Nymphaeaceae

(A) *Nymphaea alba* L.—Rooted Hydrophyte, Eurasiatic; Aquatic habitats.

Oleaceae

Fraxinus angustifolia Vahl subsp. ***oxycarpa*** (Willd.) Franco et Rocha Afonso—Phanerophyte, South-European—South-Siberian; Woodlands.

Fraxinus ornus L.—Phanerophyte, South-European—South-Siberian; Woodlands.

Ligustrum vulgare L.—Nanophanerophyte, European-Caucasian; Woodlands.

Phillyrea angustifolia L.—Phanerophyte, West-Stenomediterranean; Woodlands (Fig. 2.24).

Onagraceae

(A, P) *Epilobium hirsutum* L.—Scapose Hemicryptophyte, Palaeotemperate; Wet meadows, Ruderal habitats.

(A, P) *Epilobium parviflorum* Schreb.—Scapose Hemicryptophyte, Palaeotemperate; Wet meadows.

Fig. 2.24 *Phillyrea angustifolia* (Photo A. Alessandrini)

Fig. 2.25 *Anacamptis coriophora* (Photo R. Gerdol)

Epilobium roseum Schreb. subsp. ***roseum***—Scapose Hemicryptophyte, Eurasiatic; Wet meadows.

(A, P) *Epilobium tetragonum* L.—Scapose Hemicryptophyte, Palaeotemperate; Wet meadows, Ruderal habitats.

Oenothera stucchii Soldano—Scapose Hemicryptophyte, Invasive Alien; Ruderal habitats.

Orchidaceae

Anacamptis coriophora (L.) R.M. Bateman, Pridgeon et M.W. Chase—Bulbous Geophyte, Eurimediterranean; Dry grasslands (Fig. 2.25).

(A, P) *Anacamptis laxiflora* (Lam.) R.M. Bateman, Pridgeon et M.W. Chase—Bulbous Geophyte, Eurimediterranean; Wet meadows.

Anacamptis morio (L.) R.M. Bateman, Pridgeon et M.W. Chase subsp. ***morio***—Bulbous Geophyte, European-Caucasian; Dry grasslands.

Fig. 2.26 *Anacamptis palustris* (Photo R. Gerdol)

Anacamptis palustris (Jacq.) R.M. Bateman, Pridgeon et M.W. Chase—Bulbous Geophyte, Eurimediterranean; Wet meadows. This species is undergoing strong decline all over Italy. In the region Emilia-Romagna it only occurs in coastal areas (Fig. 2.26).

Anacamptis pyramidalis (L.) Rich.—Bulbous Geophyte, Eurimediterranean; Dry grasslands.

Cephalanthera damasonium (Mill.) Druce—Rhizome Geophyte, Eurimediterranean; Woodlands.

Cephalanthera longifolia (L.) Fritsch—Rhizome Geophyte, Eurasiatic; Woodlands (Fig. 2.27).

Cephalanthera rubra (L.) Rich.—Rhizome Geophyte, Eurasiatic; Woodlands (Fig. 2.28).

(A, P) *Epipactis helleborine* (L.) Crantz—Rhizome Geophyte, Palaeotemperate; Woodlands.

★ ***Epipactis microphylla*** (Ehrh.) Sw.—Rhizome Geophyte, European-Caucasian; Woodlands.

Limodorum abortivum (L.) Sw.—Rhizome Geophyte, Eurimediterranean; Woodlands.

Neotinea tridentata (Scop.) R.M. Bateman, Pridgeon et M.W. Chase—Bulbous Geophyte, Eurimediterranean; Dry grasslands.

Neottia nidus-avis (L.) Rich.—Rhizome Geophyte, Eurasiatic; Woodlands.

Ophrys sphegodes Mill. subsp. ***sphegodes***—Bulbous Geophyte, Eurimediterranean; Dry grasslands.

Orobanchaceae

Fig. 2.27 *Cephalanthera longifolia* (Photo L. Brancaleoni)

Fig. 2.28 *Cephalanthera rubra* (Photo A. Alessandrini)

Fig. 2.29 *Parentucellia latifolia* (Photo R. Gerdol)

(A, P) *Odontites vulgaris* Moench—Scapose Therophyte, Eurasiatic; Dry grasslands.

★ *Parentucellia latifolia* (L.) Caruel—Scapose Therophyte, Eurimediterranean; Dry grasslands (Fig. 2.29).

★ *Parentucellia viscosa* (L.) Caruel—Scapose Therophyte, Eurimediterranean-Subatlantic; Dry grasslands.

Oxalidaceae

Oxalis corniculata L.—Reptant Chamaephyte, Eurimediterranean; Ruderal habitats.

★ *Oxalis dillenii* Jacq.—Scapose Hemicryptophyte, Naturalized Alien; Ruderal habitats.

Papaveraceae

Chelidonium majus L.—Scapose Hemicryptophyte, Eurasiatic; Ruderal habitats.

(A, P) *Fumaria officinalis* L.—Scapose Therophyte, Palaeotemperate; Ruderal habitats.

(A) *Glaucium flavum* Crantz—Scapose Hemicryptophyte, Eurimediterranean; Dry grasslands.

Papaver rhoeas L.—Scapose Therophyte, East-Mediterranean-Montane; Ruderal habitats.

Phytolaccaceae

Fig. 2.30 *Gratiola officinalis* (Photo R. Gerdol)

Phytolacca americana L.—Rhizome Geophyte, Naturalized Alien; Ruderal habitats.

Pinaceae

Pinus pinaster Aiton—Phanerophyte, Naturalized Alien (Feral); Woodlands.
Pinus pinea L.—Phanerophyte, Naturalized Alien (Feral); Woodlands.

Plantaginaceae

Gratiola officinalis L.—Scapose Hemicryptophyte, Boreal; Wet meadows (Fig. 2.30).
Kickxia commutata (Bernh. ex Rchb.) Fritsch—Reptant Chamaephyte, Stenomediterranean, Ruderal habitats (Fig. 2.31).
Linaria vulgaris Mill. subsp. ***vulgaris***—Scapose Hemicryptophyte, Eurasiatic; Ruderal habitats.
★ ***Plantago afra*** L.—Scapose Therophyte, Stenomediterranean; Dry grasslands.
Plantago altissima L.—Scapose Hemicryptophyte (Rosette), Southeast-European; Wet meadows.
(A, P) *Plantago arenaria* Waldst. et Kit.—Scapose Therophyte, Southeast-European; Dry grasslands.
Plantago coronopus L.—Scapose Therophyte, Eurimediterranean; Brackish wetlands.
Plantago lanceolata L.—Scapose Hemicryptophyte (Rosette), Eurasiatic; Ruderal habitats.
Plantago major L.—Scapose Hemicryptophyte (Rosette), Eurasiatic; Ruderal habitats.
(A, P) *Plantago sempervirens* Crantz—Suffrutescent Chamaephyte, West-Stenomediterranean; Dry grasslands.
Veronica agrestis L.—Scapose Therophyte, European; Ruderal habitats.
Veronica anagallis-aquatica L. subsp. ***anagallis-aquatica***—Scapose Hemicryptophyte, Cosmopolitan; Wet meadows.
★ ***Veronica anagalloides*** Guss.—Scapose Therophyte, Eurimediterranean; Wet meadows.

Fig. 2.31 *Kickxia commutata* (Photo A. Alessandrini)

Veronica arvensis L.—Scapose Therophyte, Palaeotemperate; Ruderal habitats.
Veronica catenata Pennell—Scapose Hemicryptophyte, Boreal; Wet meadows.
★ *Veronica hederifolia* L.—Scapose Therophyte, Eurasiatic; Ruderal habitats.
Veronica officinalis L.—Reptant Chamaephyte, Eurasiatic; Woodlands.
Veronica persica Poir.—Scapose Therophyte, Invasive Alien; Ruderal habitats.

Plumbaginaceae

(A, P) *Limonium bellidifolium* (Gouan) Dumort.—Scapose Hemicryptophyte (Rosette), Eurimediterranean-Turanian; Brackish wetlands.
Limonium narbonense Mill.—Scapose Hemicryptophyte (Rosette), Eurimediterranean; Brackish wetlands.
(A, P) *Limonium virgatum* (Willd.) Fourr.—Scapose Hemicryptophyte (Rosette), Eurimediterranean; Brackish wetlands.

Poaceae

(A) *Aeluropus littoralis* (Gouan) Parl.—Rhizome Geophyte, Eurimediterranean-Turanian; Dry grasslands.
★ *Agrostis castellana* Boiss. et Reut.—Caespitose Hemicryptophyte, West-Eurimediterranean; Woodlands.

Agrostis stolonifera L. subsp. *stolonifera*—Reptant Chamaephyte, Boreal; Wet meadows, Ruderal habitats.

★ *Aira elegantissima* Schur—Scapose Therophyte, Eurimediterranean; Dry grasslands, Ruderal habitats.

Alopecurus myosuroides Huds.—Scapose Therophyte, Subcosmopolitan; Ruderal habitats.

Anthoxanthum odoratum L.—Caespitose Hemicryptophyte, Eurasiatic; Dry grasslands.

★ *Arundo donax* L.—Rhizome Geophyte, Naturalized Alien (Feral); Ruderal habitats.

★ *Avena barbata* Pott ex Link—Scapose Therophyte, Eurimediterranean; Ruderal habitats.

★ *Brachypodium rupestre* (Host) Roem. et Schult. subsp. *caespitosum* (Host) H. Scholz—Caespitose Hemicryptophyte, Subatlantic; Dry grasslands.

Brachypodium rupestre (Host) Roem. et Schult. subsp. *rupestre*—Caespitose Hemicryptophyte, Subatlantic; Dry grasslands.

Brachypodium sylvaticum (Huds.) P. Beauv.—Caespitose Hemicryptophyte, Palaeotemperate; Woodlands.

★ *Bromus diandrus* Roth [*Anisantha diandra* (Roth) Tzvelev]—Scapose Therophyte, Eurimediterranean; Ruderal habitats.

Bromus erectus Huds. subsp. *erectus*—Caespitose Hemicryptophyte, Palaeotemperate; Dry grasslands.

Bromus hordeaceus L.—Scapose Therophyte, Subcosmopolitan; Dry grasslands.

(A, P) *Bromus madritensis* L. [*Anisantha m.* (L.) Nevski]—Scapose Therophyte, Eurimediterranean; Dry grasslands.

(A, P) *Bromus ramosus* Huds.—Caespitose Hemicryptophyte, Eurasiatic; Woodlands.

Bromus squarrosus L.—Scapose Therophyte, Palaeotemperate; Ruderal habitats, Dry grasslands.

(A, P) *Bromus sterilis* L. [*Anisantha s.* (L.) Nevski]—Scapose Therophyte, Eurimediterranean; Ruderal habitats.

Bromus tectorum L. [*Anisantha t.* (L.) Nevski]—Scapose Therophyte, Palaeotemperate; Ruderal habitats.

Calamagrostis arundinacea (L.) Roth—Caespitose Hemicryptophyte, Eurasiatic; Wet meadows, Reeds.

Calamagrostis epigejos (L.) Roth—Caespitose Hemicryptophyte, Euro-Siberian; Wet meadows.

Calamagrostis pseudophragmites (Haller f.) Koeler—Caespitose Hemicryptophyte, Euro-Siberian; Wet meadows.

Catapodium rigidum (L.) C.E. Hubb.—Scapose Therophyte, Eurimediterranean; Ruderal habitats.

Cenchrus spinifex Cav.—Scapose Therophyte, Invasive Alien; Dry grasslands.

Corynephorus articulatus (Desf.) P. Beauv.—Scapose Therophyte, Stenomediterranean; Dry grasslands.

Cynodon dactylon (L.) Pers.—Rhizome Geophyte, Cosmopolitan; Ruderal habitats.

★ *Cynosurus cristatus* L.—Caespitose Hemicryptophyte, European-Caucasian; Ruderal habitats.

Dactylis glomerata L.—Caespitose Hemicryptophyte, Palaeotemperate; Dry grasslands. The Bosco della Mesola populations belong in part to the nominal subspecies and in part to subsp. *lobata* (Drejer) H. Lindb.

Danthonia decumbens (L.) DC.—Caespitose Hemicryptophyte, European; Wet meadows.

Dasypyrum villosum (L.) Borbás—Scapose Therophyte, Eurimediterranean-Turanian; Dry grasslands.

★ *Digitaria ciliaris* (Retz.) Koeler—Scapose Therophyte, Subtropical; Ruderal habitats.

Digitaria sanguinalis (L.) Scop.—Scapose Therophyte, Cosmopolitan; Ruderal habitats.

Echinochloa crus-galli (L.) P. Beauv.—Scapose Therophyte, Subcosmopolitan; Ruderal habitats.

Eleusine indica (L.) Gaertn.—Scapose Therophyte, Naturalized Alien; Ruderal habitats.

Elymus athericus (Link) Kerguélen [*Elytrigia a.* (Link) Kerguélen ex Carreras Martinez]—Rhizome Geophyte, Eurimediterranean; Wet meadows, Brackish wetlands.

(A, P) *Elymus farctus* (Viv.) Runemark ex Melderis [*Elytrigia juncea* (L.) Nevski]—Rhizome Geophyte, Eurimediterranean; Dry grasslands.

Elymus repens (L.) Gould [*Elytrigia repens* (L.) Nevski]—Rhizome Geophyte, Boreal; Wet meadows, Ruderal habitats.

★ *Festuca pratensis* Huds. [*Schedonorus pratensis* (Huds.) P. Beauv.]—Caespitose Hemicryptophyte, Eurasiatic; Dry grasslands.

(A, P) *Festuca rubra* L.—Caespitose Hemicryptophyte, Boreal; Dry grasslands.

★ *Glyceria maxima* (Hartm.) Holmb.—Rooted Hydrophyte, Boreal; Wet meadows, Reeds.

Holcus lanatus L.—Caespitose Hemicryptophyte, Boreal; Dry grasslands.

Hordeum marinum Huds.—Scapose Therophyte, West-Eurimediterranean; Dry grasslands, Ruderal habitats.

Hordeum murinum L. subsp. *leporinum* (Link) Arcang.—Scapose Therophyte, Boreal; Ruderal habitats.

(A, P) *Hordeum secalinum* Schreb.—Caespitose Hemicryptophyte, West-Eurimediterranean-Subatlantic; Ruderal habitats.

Lagurus ovatus L.—Scapose Therophyte, Eurimediterranean; Dry grasslands.

★ *Lolium multiflorum* Lam.—Scapose Therophyte, Eurimediterranean; Ruderal habitats.

Lolium perenne L.—Caespitose Hemicryptophyte, Boreal; Ruderal habitats.

Molinia arundinacea Schrank—Caespitose Hemicryptophyte, European-Caucasian; Wet meadows.

(A, P) *Panicum miliaceum* L.—Scapose Therophyte, Casual Alien (Feral); Ruderal habitats.

(A, P) *Parapholis incurva* (L.) C.E. Hubb.—Scapose Therophyte, Stenomediterranean-Atlantic; Dry grasslands, Brackish wetlands.

★ ***Paspalum distichum*** L.—Rhizome Geophyte, Invasive Alien; Wet meadows, Ruderal habitats.

Phalaris arundinacea L.—Helophyte, Boreal; Wet meadows, Reeds.

Phleum arenarium L. subsp. ***caesium*** H. Scholz—Scapose Therophyte, Stenomediterranean-Atlantic; Dry grasslands.

Phragmites australis (Cav.) Trin. ex Steud.—Rhizome Geophyte, Subcosmopolitan; Wet meadows, Reeds.

Poa annua L.—Caespitose Therophyte, Cosmopolitan; Ruderal habitats.

Poa bulbosa L.—Caespitose Hemicryptophyte, Palaeotemperate; Ruderal habitats.

Poa pratensis L.—Caespitose Hemicryptophyte, Boreal; Dry grasslands, Ruderal habitats.

Poa sylvicola Guss.—Caespitose Hemicryptophyte, Eurimediterranean; Dry grasslands, Ruderal habitats.

Poa trivialis L.—Caespitose Hemicryptophyte, Eurasiatic; Dry grasslands, Ruderal habitats.

Polypogon monspeliensis (L.) Desf.—Scapose Therophyte, Subtropical; Dry grasslands.

Puccinellia distans (Jacq.) Parl.—Caespitose Hemicryptophyte, Palaeotemperate; Brackish wetlands.

Puccinellia festuciformis (Host) Parl.—Caespitose Hemicryptophyte, Stenomediterranean; Brackish wetlands.

★ ***Rostraria cristata*** (L.) Tzvelev [*Lophochloa c.* (L.) Hyl.]—Scapose Therophyte, Subcosmopolitan; Ruderal habitats.

★ ***Rostraria hispida*** (Savi) Doğan [*Lophochloa h.* (Savi) Pignatti]—Scapose Therophyte, Stenomediterranean; Dry grasslands, Ruderal habitats.

Saccharum ravennae (L.) L. subsp. ***ravennae*** [*Tripidium ravennae* (L.) H. Scholz]—Caespitose Hemicryptophyte, Eurimediterranean Turanian; Wet meadows, Brackish wetlands (Fig. 2.32).

Scolochloa festucacea (Willd.) Link [*Schedonorus arundinaceus* (Schreb.) Dumort.]—Caespitose Hemicryptophyte, Palaeotemperate; Dry grasslands.

Setaria pumila (Poir.) Roem. et Schult.—Scapose Therophyte, Subcosmopolitan; Ruderal habitats.

★ ***Setaria verticillata*** (L.) P. Beauv.—Scapose Therophyte, Subtropical; Ruderal habitats.

Sorghum halepense (L.) Pers.—Rhizome Geophyte, Invasive Alien; Ruderal habitats.

Sporobolus anglicus (C.E. Hubb.) P.M. Peterson et Saarela—Rhizome Geophyte, Invasive Alien; Brackish wetlands.

Sporobolus maritimus P.M. Peterson et Saarela—Rhizome Geophyte, Euro-American; Brackish wetlands.

(A, P) *Tragus racemosus* (L.) All.—Scapose Therophyte, Cosmopolitan; Dry grasslands, Ruderal habitats.

Fig. 2.32 *Saccharum ravennae* subsp. *ravennae* (Photo A. Alessandrini)

★ ***Trisetaria flavescens*** (L.) Baumg.—Caespitose Hemicryptophyte, Eurasiatic; Dry grasslands, Ruderal habitats.

Trisetaria michelii (Savi) D. Heller—Scapose Therophyte, Stenomediterranean; Dry grasslands.

Vulpia ciliata Dumort.—Caespitose Therophyte, Eurimediterranean; Ruderal habitats.

Vulpia fasciculata (Forssk.) Fritsch—Caespitose Therophyte, Eurimediterranean-Subatlantic; Dry grasslands.

(A, P) *Vulpia myuros* (L.) C.C. Gmel.—Caespitose Therophyte, Subcosmopolitan; Dry grasslands.

Polygalaceae

Polygala comosa Schkuhr—Scapose Hemicryptophyte, South-European—South-Siberian; Dry grasslands.

★ ***Polygala monspeliaca*** L.—Scapose Therophyte, Stenomediterranean; Dry grasslands.

Polygala nicaeensis W.D.J. Koch subsp. ***mediterranea*** Chodat—Scapose Hemicryptophyte, Eurimediterranean; Dry grasslands.

(A, P) *Polygala vulgaris* L.—Scapose Hemicryptophyte, Eurasiatic; Dry grasslands.

Polygonaceae

Fallopia convolvulus (L.) Á. Löve—Scapose Therophyte, Boreal; Ruderal habitats.

(A, P) *Fallopia dumetorum* (L.) Holub—Scapose Therophyte, Euro-Siberian; Ruderal habitats.

Persicaria lapathifolia (L.) Delarbre—Scapose Therophyte, Palaeotemperate; Wet meadows, Ruderal habitats.

(A) *Persicaria maculosa* (L.) Gray—Scapose Therophyte, Subcosmopolitan; Wet meadows, Ruderal habitats.

★ ***Persicaria minor*** (Huds.) Opiz—Scapose Therophyte, Subcosmopolitan; Wet meadows, Ruderal habitats.

★ ***Polygonum arenastrum*** Boreau—Reptant Therophyte, Subcosmopolitan; Ruderal habitats.

Polygonum aviculare L.—Reptant Therophyte, Cosmopolitan; Ruderal habitats.

(A, P) *Rumex crispus* L.—Scapose Hemicryptophyte, Subcosmopolitan; Ruderal habitats.

★ ***Rumex cristatus*** DC. subsp. ***cristatus***—Scapose Hemicryptophyte, Invasive Alien; Ruderal habitats.

★ ***Rumex obtusifolius*** L. subsp. ***obtusifolius***—Scapose Hemicryptophyte, European-Caucasian; Ruderal habitats.

Polypodiaceae

★ ***Polypodium vulgare*** (gr.)—Scapose Hemicryptophyte (Rosette), Boreal, Woodlands. This is the first record of *Polypodium vulgare* s.l. for the regional coastal area. The species identity of the Bosco della Mesola population has to be ascertained further.

Portulacaceae

Portulaca oleracea L.—Scapose Therophyte, Subcosmopolitan; Ruderal habitats.

Potamogetonaceae

Potamogeton crispus L.—Rooted Hydrophyte, Subcosmopolitan; Aquatic habitats.

Potamogeton nodosus Poir.—Rooted Hydrophyte, Subcosmopolitan; Aquatic habitats.

Potamogeton pusillus L.—Rooted Hydrophyte, Subcosmopolitan; Aquatic habitats.

Stuckenia pectinata (L.) Börner (*Potamogeton p.* L.)—Rooted Hydrophyte, Subcosmopolitan; Aquatic habitats.

Zannichellia palustris L. subsp. ***pedicellata*** (Wahlenb. et Rosén) Hook. f.—Rooted Hydrophyte, Cosmopolitan; Aquatic habitats.

Primulaceae

Anagallis arvensis L.—Reptant Therophyte, Eurimediterranean; Ruderal habitats.

★ ***Anagallis minima*** L.—Scapose Therophyte, Palaeotemperate; Wet meadows. This species is becoming more and more rare in Italy. It has been ascertained at a few stations in the region Emilia-Romagna. This is the first record in the province of Ferrara.

Lysimachia vulgaris L.—Scapose Hemicryptophyte, Eurasiatic; Ruderal habitats.

Samolus valerandi L.—Caespitose Hemicryptophyte, Cosmopolitan; Wet meadows.

Ranunculaceae

Clematis flammula L.—Liana, Eurimediterranean; Woodlands (Fig. 2.33).

Clematis vitalba L.—Liana, European-Caucasian; Woodlands.

Ficaria verna Huds. subsp. ***fertilis*** (A.R. Clapham ex Laegaard) Stace—Bulbous Geophyte, Eurasiatic; Ruderal habitats.

(A, P) *Ranunculus acris* L.—Scapose Hemicryptophyte, Subcosmopolitan; Dry grasslands.

Ranunculus neapolitanus Ten.—Scapose Hemicryptophyte, Northeast-Mediterranean-Montane; Dry grasslands, Ruderal habitats.

Ranunculus peltatus Schrank subsp. ***baudotii*** (Godr.) C.D.K. Cook—Rooted Hydrophyte, Eurimediterranean-Subatlantic; Aquatic habitats (Fig. 2.34).

Ranunculus repens L.—Reptant Chamaephyte, Palaeotemperate; Wet meadows.

★ ***Ranunculus sardous*** Crantz—Scapose Therophyte, Eurimediterranean; Wet meadows, Ruderal habitats.

Ranunculus sceleratus L.—Scapose Therophyte, Palaeotemperate; Wet meadows.

★ ***Ranunculus serpens*** Schrank subsp. ***polyanthemophyllus*** (W. Koch et H.E. Hess) Kerguélen—Scapose Hemicryptophyte, North-Eurimediterranean; Dry grasslands. Extremely rare in the region Emilia-Romagna. This is the first record in the Province of Ferrara.

★ ***Ranunculus trichophyllus*** Chaix—Rooted Hydrophyte, European; Aquatic habitats (Fig. 2.35).

(A) *Thalictrum lucidum* L.—Scapose Hemicryptophyte, Southeast-European; Reeds.

Rhamnaceae

Fig. 2.33 *Clematis flammula* (Photo L. Brancaleoni)

Frangula alnus Mill. subsp. ***alnus***—Phanerophyte, European-Caucasian; Woodlands.

Paliurus spina-christi Mill.—Phanerophyte, Southeast-European; Ruderal habitats.

Rhamnus cathartica L.—Phanerophyte, South-European—South-Siberian; Woodlands.

Rosaceae

Agrimonia eupatoria L.—Scapose Hemicryptophyte, Subcosmopolitan; Ruderal habitats.

Aphanes arvensis L.—Scapose Therophyte, Subcosmopolitan; Dry grasslands.

★ ***Aphanes australis*** Rydb.—Scapose Therophyte, Subatlantic; Dry grasslands. First record for the region Emilia-Romagna (Alessandrini et al. 2019).

Crataegus monogyna Jacq.—Phanerophyte, Palaeotemperate; Woodlands.

Malus sylvestris (L.) Mill.—Phanerophyte, Central-European; Woodlands.

(A, P) *Potentilla erecta* (L.) Räusch.—Scapose Hemicryptophyte, Eurasiatic; Woodlands, Wet meadows.

(A, P) *Potentilla pusilla* Host—Scapose Hemicryptophyte, Central-European; Woodlands, Wet meadows.

Potentilla reptans L.—Scapose Hemicryptophyte (Rosette), Palaeotemperate; Wet meadows, Ruderal habitats.

Fig. 2.34 *Ranunculus peltatus* subsp. *baudotii* (Photo L. Brancaleoni)

Potentilla tabernaemontani Asch.—Scapose Hemicryptophyte, European; Dry grasslands.

Prunus spinosa L. subsp. ***spinosa***—Phanerophyte, European-Caucasian; Woodlands, Ruderal habitats.

Pyracantha coccinea M. Roem.—Phanerophyte, Stenomediterranean; Woodlands.

Pyrus communis L. subsp. ***pyraster*** (L.) Ehrh.—Phanerophyte, Eurasiatic; Woodlands.

Rosa canina L.—Nanophanerophyte, Palaeotemperate; Woodlands, Dry grasslands.

Rosa sempervirens L.—Nanophanerophyte, Stenomediterranean; Woodlands.

Rubus caesius L.—Nanophanerophyte, Eurasiatic; Woodlands, Ruderal habitats.

Rubus ulmifolius Schott—Nanophanerophyte, Eurimediterranean; Ruderal habitats.

Sanguisorba minor Scop.—Scapose Hemicryptophyte, Palaeotemperate; Dry grasslands.

Rubiaceae

Asperula cynanchica L.—Scapose Hemicryptophyte, Eurimediterranean; Woodlands, Dry grasslands.

Galium album Mill. (*G. mollugo* L. subsp. *erectum* Syme)—Scapose Hemicryptophyte, Eurimediterranean; Ruderal habitats.

Galium aparine L.—Scapose Therophyte, Eurasiatic; Ruderal habitats.

Fig. 2.35 *Ranunculus trichophyllus* (Photo L. Brancaleoni)

Galium debile Desv.—Scapose Hemicryptophyte, Eurimediterranean; Wet meadows.
Galium palustre L.—Scapose Hemicryptophyte, European-Caucasian; Wet meadows.
Galium verum L.—Scapose Therophyte, European-Caucasian; Dry grasslands.
Rubia peregrina L.—Liana, Stenomediterranean; Woodlands.
(A) *Rubia tinctoria* L.—Scapose Hemicryptophyte, Casual Alien (Feral); Ruderal habitats.
Sherardia arvensis L.—Scapose Therophyte, Eurimediterranean; Ruderal habitats.

Ruppiaceae

Ruppia cirrhosa (Petagna) Grande—Rooted Hydrophyte, Cosmopolitan; Brackish wetlands.

Salicaceae

Populus alba L.—Phanerophyte, Palaeotemperate; Woodlands.
(A, P) *Populus nigra* L.—Phanerophyte, Palaeotemperate; Woodlands.
Populus × ***canadensis*** Moench—Phanerophyte, Naturalized Alien (Feral); Woodlands, Ruderal habitats.
Salix alba L.—Phanerophyte, Palaeotemperate; Woodlands.

Salvinlaceae

(A) *Azolla filiculoides* Lam.—Errant Hydrophyte, Naturalized Alien; Aquatic habitats.

Salvinia natans (L.) All.—Errant Hydrophyte, Temperate-Eurasiatic; Aquatic habitats.

Santalaceae

Osyris alba L.—Nanophanerophyte, Eurimediterranean; Woodlands.

Sapindaceae

Acer campestre L.—Phanerophyte, European-Caucasian; Woodlands.

★ *Acer negundo* L.—Phanerophyte, Naturalized Alien; Ruderal habitats.

(A, P) *Acer platanoides* L.—Phanerophyte, Casual Alien (Feral); Ruderal habitats.

Saxifragaceae

Saxifraga tridactylites L.—Scapose Therophyte, Eurimediterranean; Dry grasslands, Ruderal habitats.

Scrophulariaceae

(A, P) *Scrophularia canina* L.—Scapose Hemicryptophyte, Eurimediterranean; Ruderal habitats.

Verbascum blattaria L.—Scapose Hemicryptophyte, Palaeotemperate; Ruderal habitats.

★ *Verbascum densiflorum* Bertol.—Scapose Hemicryptophyte, North-Eurimediterranean; Ruderal habitats.

(A) *Verbascum phlomoides* L.—Scapose Hemicryptophyte, Eurimediterranean; Ruderal habitats.

Verbascum sinuatum L.—Scapose Hemicryptophyte, Eurimediterranean; Ruderal habitats.

(A, P) *Verbascum thapsus* L.—Scapose Hemicryptophyte, European-Caucasian; Ruderal habitats.

Simaroubaceae

Ailanthus altissima (Mill.) Swingle—Phanerophyte, Invasive Alien (Feral); Ruderal habitats.

Solanaceae

Datura stramonium L. subsp. *stramonium*—Scapose Therophyte, Invasive Alien; Ruderal habitats.

(A, P) *Hyoscyamus albus* L.—Scapose Therophyte, Eurimediterranean; Ruderal habitats.

Solanum dulcamara L.—Nanophanerophyte, Palaeotemperate; Woodlands.

Solanum nigrum L.—Scapose Therophyte, Cosmopolitan; Ruderal habitats.

Tamaricaceae

Tamarix gallica L.—Phanerophyte, West-Stenomediterranean; Brackish wetlands.

Thelypteridaceae

Thelypteris palustris Schott—Rhizome Geophyte, Subcosmopolitan; Wet meadows (Fig. 2.36).

Typhaceae

Typha angustifolia L.—Rhizome Geophyte, Boreal; Reeds.
Typha latifolia L.—Rhizome Geophyte, Cosmopolitan; Reeds.

Ulmaceae

Ulmus minor Mill.—Phanerophyte, European-Caucasian; Woodlands, Ruderal habitats.

Urticaceae

★ *Parietaria judaica* L.—Scapose Hemicryptophyte, Eurimediterranean-Macaronesian; Ruderal habitats.
Urtica dioica L. subsp. *dioica*—Scapose Hemicryptophyte, Subcosmopolitan; Ruderal habitats.
(A) *Urtica membranacea* Poir. ex Savigny—Scapose Therophyte, South-Stenomediterranean; Ruderal habitats.
Urtica urens L.—Scapose Therophyte, Subcosmopolitan; Ruderal habitats.

Verbenaceae

Verbena officinalis L. Scapose Hemicryptophyte, Palaeotemperate; Ruderal habitats.

Viburnaceae

Sambucus ebulus L.—Rhizome Geophyte, Eurimediterranean; Ruderal habitats.
Sambucus nigra L.—Phanerophyte, European-Caucasian; Ruderal habitats.
Viburnum lantana L.—Phanerophyte, Central-European; Woodlands.
Viburnum opulus L.—Phanerophyte, Eurasiatic; Woodlands.

Violaceae

★ *Viola alba* Besser subsp. *alba*—Scapose Hemicryptophyte (Rosette), Eurimediterranean; Woodlands (Fig. 2.37).
Viola alba Besser subsp. *dehnhardtii* (Ten.) W. Becker—Scapose Hemicryptophyte (Rosette), Eurimediterranean; Woodlands.
Viola canina L.—Scapose Hemicryptophyte, Eurasiatic; Woodlands.
Viola hirta L.—Scapose Hemicryptophyte (Rosette), European; Woodlands (Fig. 2.38).
Viola odorata L.—Scapose Hemicryptophyte (Rosette), Eurimediterranean; Woodlands, Ruderal habitats.
Viola suavis M. Bieb. subsp. *suavis*—Scapose Hemicryptophyte (Rosette), South-European—South-Siberian; Woodlands.

Fig. 2.36 *Thelypteris palustris* (Photo L. Brancaleoni)

Fig. 2.37 *Viola alba* subsp. *alba* (Photo M. Pellizzari)

Fig. 2.38 *Viola hirta* (Photo M. Pellizzari)

Fig. 2.39 *Tribulus terrestris* (Photo A. Alessandrini)

Vitaceae

★ *Vitis* × *koberi* Ardenghi, Galasso, Banfi et Lastrucci—Liana, Invasive Alien (Feral); Ruderal habitats.
Vitis vinifera L.—Liana, Naturalized Alien (Feral); Ruderal habitats.

Zygophyllaceae

Tribulus terrestris L.—Reptant Therophyte, Cosmopolitan; Ruderal habitats (Fig. 2.39).

2.3 Synthesis and Conclusion

The vascular flora of the Bosco della Mesola Nature Reserve is currently comprised of 480 taxa, i.e. *ca.* 25% more than those recorded in the previous checklist dating back to the early 1960s (Piccoli 1964). Such increase was principally caused by arrival of new species. These are mostly native species (such as *Carduus pycnocephalus* subsp. *pycnocephalus*, *Silybum marianum*, *Tragopogon porrifolius*, *Capsella rubella*, *Minuartia mediterranea*, *Polycarpon tetraphyllum*, *Stellaria pallida*, *Aira elegantissima*, *Bromus diandrus*, *Digitaria ciliaris*, *Rostraria cristata*, *Polygonum arenastrum*, *Ranunculus sardous* and others) or, to a lesser extent, aliens (such as *Crepis sancta*, *Symphyotrichum squamatum*, *Commelina communis*, *Dichondra micrantha*, *Elaeagnus angustifolia*, *Amorpha fruticosa*, *Oxalis dillenii*, *Arundo donax*, *Paspalum distichum* and others). Arrival of new species took place principally in ruderal habitats. Over 60 years, the frequency of species preferentially occurring in ruderal habitats increased by 5.4% (Table 2.1) while the frequency of

Table 2.1 Percentage frequencies of species grouped according to habitat preference in the Piccoli's checklist (1960) and in the current checklist (2020)

	1960	2020
Woodlands	14.3	14.0
Dry grasslands	23.9	18.8
Wet meadows	14.8	15.8
Reeds	3.2	4.4
Brackish wetlands	7.0	5.8
Aquatic habitats	4.1	3.1
Ruderal habitats	32.7	38.1
Total	100	100

Table 2.2 Percentage frequencies of species grouped according to life forms in the Piccoli's checklist (1960) and in the current checklist (2020)

	1960	2020
Chamaephytes	6.0	5.0
Geophytes	8.9	12.5
Helophytes	3.4	4.2
Hemicryptophytes	35.8	31.3
Hydrophytes	5.0	3.8
Lianas	1.6	1.9
Phanerophytes	10.2	11.0
Therophytes	29.2	30.4
Total	100	100

species preferentially occurring in dry grasslands declined by almost the same order of magnitude. Species frequencies in the other habitat types underwent much smaller changes (Table 2.1). The records of more or less rare species not reported in the previous checklist (*Oenanthe lachenalii, Achillea nobilis, Helianthemum apenninum, Carex punctata, Epipactis microphylla, Parentucellia latifolia, Parentucellia viscosa, Agrostis castellana, Polypodium vulgare, Anagallis minima* and *Ranunculus serpens* subsp. *polyanthemophyllus*) is probably due to the detailed field surveys carried out in the last three decades. The records of *Achillea roseoalba, Brachypodium rupestre* and *Aphanes australis* are due to deepening of taxonomic research in recent years. Only in quite a few cases were some rather common species presumably overlooked in Piccoli's (1964) surveys (this may be the case, for example, of *Stachys palustris, Festuca pratensis* and *Glyceria maxima*). Other species, documented only in the early 1900s, have certainly disappeared for decades: *Agrostemma gitago, Apocynum venetum, Glaucium flavum, Ambrosia maritima, Triglochin maritimum, Nymphaea alba, Anacamptis laxiflora*.

About two thirds of the Bosco della Mesola flora is comprised of herbaceous terrestrial species (hemicryptophytes + therophytes + geophytes). Hemicryptophytes decreased by 4.5% while geophytes increased by 3.6% over 60 years. The frequencies of the other life forms varied little (Table 2.2). With respect to chorotypes, Mediterranean and Eurasiatic species make up almost half of the Bosco della Mesola flora (Table 2.3). Mediterranean species increased by 3.5%, while temperate species and cosmopolitan species decreased by *ca.* 2.5% each over 60 years (Table 2.3). The frequencies of alien species increased from less than 6% to *ca.* 8%. The frequencies

Table 2.3 Percentage frequencies of species grouped according to main chorotypes in the Piccoli's checklist (1960) and in the current checklist (2020)

	1960	2020
Mediterranean	23.2	26.7
Eurasiatic	22.5	22.1
Temperate	15.4	12.3
Cosmopolitan	14.6	12.1
SE European—W Asiatic	8.6	7.5
Boreal	5.5	6.5
Atlantic	3.7	3.5
Tropical	0.5	1.0
Endemic	0.3	0.2
Alien	5.7	8.1
Total	100	100

Table 2.4 Mean Ellenberg ecological indicator values in the Piccoli's checklist (1960) and in the current checklist (2020)

	1960	2020
Light value (L)	7.63	7.50
Temperature value (T)	6.71	6.71
Continentality value (C)	4.88	4.85
Water value (U)	5.09	5.20
Reaction value (R)	6.34	6.20
Nitrogen value (N)	4.39	4.59

of the other chorotypes remained almost identical over time (Table 2.3). The Ellenberg indicator values (Table 2.4) suggest that the Bosco della Mesola flora underwent very small, if any, changes in terms of species' ecological requirements over 60 years. Indeed, the average of most indicator values in 2020 was almost equal to those in 1960 with only the nitrogen indicator value experiencing a slight increase by *ca.* 5% (Table 2.4).

The Bosco della Mesola hosts more than one third of the vascular flora recorded in the Province of Ferrara, with species density of 0.45 species per ha vs. 0.005 species per ha over the entire provincial territory (Piccoli et al. 2014). About 12% of the Bosco della Mesola vascular flora is comprised of protected species at different territorial rank. Most of the protected species are listed in the Regional Law and/or in the Regional Red List (Table 2.5). This is the case of all species in the Orchidaceae family as well as some locally rare species such as *Bassia crassifolia*, *Oenanthe lachenalii*, *Cladium mariscus*, *Najas minor* and *Juncus subnodulosus*. Of even higher conservation value are *Hydrocotyle vulgaris*, *Carex rostrata*, *Euphorbia lucida*, *Utricularia australis* and *Ranunculus peltatus* subsp. *baudotii*, all listed in the Italian Red List (Table 2.5). *Kosteletzkya pentacarpos* deserves top conservation interest for being classified as 'Vulnerable' in Europe. Accordingly, this species is listed both in the 92/43/CEE Habitats Directive (Annex II) and in the Berne Convention (Table 2.5).

In spite of the environmental changes experienced in the last decades, with particular regard to climate warming (Fig. 1.2), the Bosco della Mesola underwent overall modest changes. The average ecological requirements of the vascular species

Table 2.5 List of *taxa* of conservation interest

	Regional Law	Regional Red List[a]	National Red List[b]	International convention
Alismataceae				
Alisma lanceolatum		NT		
Amaranthaceae				
Bassia crassifolia		CR		
Salicornia patula		EN		
Apiaceae				
Oenanthe lachenalii		EN		
Araceae				
Lemna gibba		VU		
Lemna minor		VU		
Lemna trisulca		VU		
Araliaceae				
Hydrocotyle vulgaris		CR	EN	
Asparagaceae				
Ruscus aculeatus		NT		Habitat Directive 92/43/CEE
Asteraceae				
Centaurea tommasinii		CR	VU	
Cirsium vulgare			DD	
Betulaceae				
Carpinus orientalis subsp. *orientalis*				Rete Natura 2000 MGC
Ceratophyllaceae				
Ceratophyllum demersum		EN		
Cistaceae				
Cistus creticus subsp. *eriocephalus*	L.R. 2/1977			
Cyperaceae				
Carex rostrata		NT	NT	
Cladium mariscus		EN		
Schoenoplectus lacustris		NT		
Schoenoplectus mucronatus		EN		
Schoenoplectus tabernaemontani		VU		
Schoenus nigricans		EN		
Euphorbiaceae				
Euphorbia lucida		LC	NT	

(continued)

Table 2.5 (continued)

	Regional Law	Regional Red List[a]	National Red List[b]	International convention
Euphorbia palustris		EN	VU	
Haloragaceae				
Myriophyllum spicatum		VU		
Hydrocharitaceae				
Najas minor		EN		
Juncaceae				
Juncus subnodulosus		CR		
Juncus tenageja		CR		
Lentibulariaceae				
Utricularia australis		DD	NT	
Linaceae				
Linum maritimum				Rete Natura 2000 MGC
Lythraceae				
Lythrum hyssopifolia			LC	
Malvaceae				
Kosteletzkya pentacarpos				Habitat Directive 92/43/CEE and Berne Convention; Rete Natura 2000 MGC
Orchidaceae				
Anacamptis coriophora	L.R. 2/1977			
Anacamptis morio subsp. *morio*	L.R. 2/1977			
Anacamptis palustris	L.R. 2/1977	EN		
Anacamptis pyramidalis	L.R. 2/1977	LC		
Cephalanthera damasonium	L.R. 2/1977			
Cephalanthera longifolia	L.R. 2/1977			
Cephalanthera rubra	L.R. 2/1977			
Epipactis microphylla	L.R. 2/1977			
Limodorum abortivum	L.R. 2/1977			
Neotinea tridentata	L.R. 2/1977			

(continued)

Table 2.5 (continued)

	Regional Law	Regional Red List[a]	National Red List[b]	International convention
Neottia nidus-avis	L.R. 2/1977			
Ophrys sphegodes subsp. *sphegodes*	L.R. 2/1977			
Plantaginaceae				
Veronica anagalloides		EN		
Plumbaginaceae				
Limonium narbonense	D.P.G. 664/1989	EN		
Poaceae				
Puccinellia distans		NT		
Puccinellia festuciformis		VU		
Saccharum ravennae subsp. *ravennae*		EN		
Sporobolus maritimus			EN	
Potamogetonaceae				
Potamogeton pusillus		VU		
Zannichellia palustris subsp. *pedicellata*		EN	DD	
Primulaceae				
Samolus valerandi		EN	LC	
Ranunculaceae				
Ranunculus peltatus subsp. *baudotii*		EN	NT	
Ruppiaceae				
Ruppia cirrhosa		EN	NT	
Salviniaceae				
Salvinia natans		EN		Rete Natura 2000 MGC
Thelypteridaceae				
Thelypteris palustris		EN	VU	
Typhaceae				
Typha angustifolia		NT		
Typha latifolia		LC		

[a] https://ambiente.regione.emilia-romagna.it/it/parchi-natura2000/sistema-regionale/flora
[b] Orsenigo et al. (2020)

pool were almost totally unaffected by climate warming. In particular, there was no sign of floristic thermophilization (sensu Gottfried et al. 2012) as demonstrated by the temperature indicator value remaining identical for about 60 years (Table 2.4). Neither did the water indicator value reveal any sign of increased frequency of drought-tolerant species. Only the nitrogen indicator value did increase across the 60-year period, although to a very slight extent, indicating somewhat higher frequency of nitrophilous species. This can be due to: increased nitrogen influx from nearby agricultural areas (Gerdol et al. 2018), manuring effect of higher density of deer in ruderal habitats, or a combination of these two causes. The chorological features of the Bosco della Mesola flora also underwent some changes especially because of a somewhat increased frequency of alien species (Table 2.3). Nonetheless, the frequency of non-native species (*ca.* 8%) still is less than half of that in the regional flora (19%; Galasso et al. 2018).

In conclusion, the Bosco della Mesola Nature Reserve plays an important role in preserving the floristic heritage, in spite of the strong climatic changes it has experienced for at least three decades. Indeed, this protected area acts as a buffer against plant diversity loss and invasion of alien species. On the other hand, future management plans must include measures to contain the autochtonous deer population by translocation of selected individuals to suitable nearby sites and reduce eutrophication in order to guarantee effective protection of the important floristic heritage of this area.

Appendix A: List of the Ellenberg's indicator values. X = undetermined

Species	L	T	C	U	R	N
Alismataceae						
Alisma lanceolatum	7	7	5	10	6	7
Alisma plantago-aquatica	7	X	X	10	X	8
Amaranthaceae						
Amaranthus albus	9	9	6	3	X	7
Amaranthus deflexus	8	8	5	4	6	9
Amaranthus retroflexus	9	9	7	4	X	9
Arthrocnemum macrostachyum	11	9	4	8	9	7
Atriplex portulacoides	11	9	4	2	6	7
Atriplex prostrata	9	X	X	6	X	9
Atriplex rosea	9	9	7	2	6	1
Atriplex tatarica	9	7	8	3	X	6
Bassia crassifolia	9	6	6	8	X	8
Bassia laniflora	9	9	6	2	5	1
Chenopodium album	7	7	5	4	5	7
Cycloloma atriplicifolium	9	7	7	4	6	1
Dysphania ambrosioides	8	7	5	2	5	5
Salicornia patula	11	7	X	8	8	7
Salsola soda	9	9	5	8	9	7
Salsola tragus	9	7	8	8	7	8
Sarcocornia fruticosa	11	9	5	8	9	7
Suaeda maritima	9	6	2	8	7	7
Amaryllidaceae						
Allium vineale	8	7	5	4	X	7
Apiaceae						
Anthriscus caucalis	7	8	5	4	6	4
Apium graveolens	7	7	5	7	5	7
Berula erecta	8	6	4	10	X	7

(continued)

A. Alessandrini et al., *The Vascular Flora of the Bosco della Mesola Nature Reserve (Northern Italy)*, Geobotany Studies, https://doi.org/10.1007/978-3-030-63412-4

Species	L	T	C	U	R	N
Bupleurum tenuissimum	11	8	5	4	7	2
Daucus carota	8	6	5	4	5	4
Echinophora spinosa	12	8	5	4	7	1
Eryngium maritimum	11	8	3	4	7	1
Oenanthe lachenalii	9	8	3	7	7	4
Oenanthe pimpinelloides	5	7	3	4	5	4
Suaeda vera	11	10	5	8	9	7
Thysselinum palustre	7	5	6	9	X	4
Torilis nodosa	7	8	6	4	7	6
Apocynaceae						
Apocynum venetum	9	8	6	2	7	1
Vincetoxicum hirundinaria subsp. *hirundinaria*	6	5	5	3	7	3
Araceae						
Lemna gibba	7	6	5	12	7	8
Lemna minor	7	X	5	12	X	X
Lemna minuta	7	X	5	12	X	X
Lemna trisulca	8	X	5	12	7	6
Spirodela polyrhiza	7	6	5	12	X	7
Araliaceae						
Hedera helix	4	5	4	5	X	X
Hydrocotyle vulgaris	9	6	5	9	3	3
Aristolochiaceae						
Aristolochia clematitis	6	7	5	4	8	8
Aristolochia rotunda	6	7	5	4	6	3
Asparagaceae						
Asparagus acutifolius	6	9	4	2	5	5
Asparagus officinalis	8	8	5	5	5	5
Ruscus aculeatus	4	8	5	4	5	5
Asteraceae						
Achillea collina	9	6	6	2	7	2
Achillea nobilis	8	7	7	4	8	1
Achillea roseoalba	7	6	7	3	X	3
Ambrosia maritima	11	8	5	1	X	1
Ambrosia psilostachya	9	7	6	2	X	1
Artemisia alba	9	5	7	3	7	2
Artemisia caerulescens subsp. *caerulescens*	11	7	5	5	9	2
Artemisia campestris	9	6	5	3	5	2
Artemisia vulgaris	9	7	8	4	X	5
Bellis perennis	9	5	4	X	X	5
Bidens tripartitus subsp. *tripartitus*	8	X	X	8	X	8
Carduus acanthoides	9	5	6	3	X	8
Carduus nutans subsp. *nutans*	8	X	5	3	8	6
Carduus pycnocephalus subsp. *pycnocephalus*	7	8	4	3	X	3

(continued)

Species	L	T	C	U	R	N
Centaurea calcitrapa	11	9	5	3	X	5
Centaurea nigrescens	7	6	5	4	5	4
Centaurea tommasinii	9	7	4	2	7	1
Chondrilla juncea	8	7	5	3	8	X
Cichorium intybus	9	6	5	3	8	5
Cirsium arvense	8	X	X	4	X	7
Cirsium vulgare	8	5	5	5	X	8
Crepis foetida	11	9	5	2	X	2
Crepis neglecta	7	6	3	4	6	3
Crepis pulchra	8	6	5	4	5	5
Crepis sancta subsp. *nemausensis*	11	9	6	2	X	2
Crepis setosa	11	9	6	2	8	2
Crepis vesicaria subsp. *taraxacifolia*	8	8	3	3	6	2
Dittrichia viscosa subsp. *viscosa*	11	8	5	3	7	9
Erigeron canadensis	8	6	5	5	X	7
Eupatorium cannabinum	7	7	5	7	5	7
Galactites tomentosus	8	8	4	3	X	7
Helichrysum italicum	8	8	5	4	3	2
Helminthotheca echioides	11	8	5	2	X	2
Hypochaeris glabra	11	8	5	2	2	1
Hypochaeris radicata	9	8	4	2	X	1
Inula salicina	7	5	5	4	9	2
Jacobaea erratica	7	6	5	8	4	5
Lactuca saligna	11	7	7	4	6	4
Lactuca serriola	9	7	7	4	6	4
Laphangium luteoalbum	7	6	5	7	5	3
Lapsana communis subsp. *communis*	5	X	5	5	X	7
Leontodon hispidus	8	X	4	4	X	3
Leontodon saxatilis subsp. *saxatilis*	11	6	5	5	7	2
Limbarda crithmoides	6	6	4	4	7	3
Matricaria chamomilla	7	5	5	6	5	5
Onopordum acanthium	11	7	6	4	7	8
Picris hieracioides	8	X	5	4	8	4
Pilosella officinarum	8	X	4	3	4	2
Pilosella piloselloides	8	6	6	3	7	2
Pulicaria dysenterica	8	6	5	7	X	5
Pulicaria vulgaris	7	7	5	7	7	7
Scolymus hispanicus subsp. *hispanicus*	11	8	5	3	X	2
Senecio inaequidens	9	7	5	2	5	1
Senecio vulgaris	7	X	X	5	X	8
Silybum marianum	11	10	6	3	5	7
Solidago canadensis	8	X	5	6	X	7
Solidago gigantea	8	X	5	X	X	7

(continued)

 Appendix A: List of the Ellenberg's indicator values. X = undetermined

Species	L	T	C	U	R	N
Sonchus asper	7	5	X	4	7	7
Sonchus maritimus	11	8	5	5	9	3
Sonchus oleraceus	7	5	X	4	8	8
Symphyotrichum squamatum	8	8	5	4	7	7
Taraxacum fulvum (gr.)	8	7	5	3	8	X
Taraxacum officinale (gr.)	7	X	X	5	X	7
Taraxacum palustre (gr.)	8	X	5	8	8	0
Tragopogon orientalis	7	5	4	4	7	5
Tragopogon porrifolius	9	9	5	3	5	3
Tragopogon pratensis	7	5	4	4	7	5
Tripolium pannonicum	8	7	X	9	7	7
Tussilago farfara	8	X	5	6	8	7
Xanthium orientale subsp. *italicum*	8	7	5	5	X	1
Xanthium spinosum	9	10	5	2	X	1
Xanthium strumarium	8	7	5	5	X	6
Berberidaceae						
Berberis vulgaris	6	6	5	4	8	3
Betulaceae						
Alnus glutinosa	5	5	5	9	6	8
Carpinus betulus	4	6	4	X	X	X
Carpinus orientalis subsp. *orientalis*	4	7	6	3	4	5
Boraginaceae						
Anchusa azurea	11	8	5	3	4	4
Anchusa officinalis	9	8	6	3	7	5
Buglossoides purpurocaerulea	5	7	6	4	8	4
Cynoglossum creticum	11	9	5	3	X	7
Cynoglossum officinale	8	5	5	3	7	8
Echium vulgare	9	7	5	4	5	4
Heliotropium europaeum	11	8	5	3	7	2
Lithospermum officinale	6	X	5	X	8	6
Myosotis arvensis	6	5	5	5	X	6
Myosotis ramosissima	9	8	5	2	4	3
Symphytum officinale	7	6	4	8	X	8
Brassicaceae						
Alyssum alyssoides	11	6	5	3	8	1
Arabidopsis thaliana	6	X	5	4	5	4
Arabis hirsuta	7	5	5	4	8	X
Arabis sagittata	7	6	6	4	8	3
Arabis turrita	6	7	6	6	7	3
Cakile maritima subsp. *maritima*	9	8	2	6	X	8
Capsella bursa-pastoris	7	X	5	5	5	4
Capsella rubella	8	9	5	2	4	2
Cardamine hirsuta	7	8	5	3	5	4

(continued)

Species	L	T	C	U	R	N
Cardamine pratensis	5	5	X	7	X	X
Cardaria draba	8	7	7	3	8	4
Descurainia sophia	7	7	8	3	6	2
Diplotaxis tenuifolia	8	7	5	4	6	5
Erophila verna	9	7	4	2	4	1
Eruca vesicaria	7	8	6	3	5	5
Hornungia petraea	9	7	5	2	6	2
Raphanus raphanistrum	11	5	5	X	4	5
Rapistrum rugosum	7	7	5	4	5	5
Sisymbrium officinale	8	6	5	4	X	7
Cannabaceae						
Humulus lupulus	7	6	4	8	6	8
Caprifoliaceae						
Dipsacus fullonum	6	8	5	7	5	5
Knautia arvensis	7	5	5	4	5	3
Lomelosia argentea	9	8	6	2	7	2
Lonicera caprifolium	6	5	6	6	X	5
Lonicera etrusca	7	8	5	3	6	4
Scabiosa triandra	7	5	7	3	5	2
Valeriana officinalis	7	6	5	8	7	5
Valerianella locusta	7	5	5	5	7	X
Caryophyllaceae						
Agrostemma githago	7	X	X	X	4	3
Arenaria leptoclados	9	9	5	2	3	1
Arenaria serpyllifolia subsp. *serpyllifolia*	9	5	X	4	X	X
Cerastium glomeratum	7	X	5	5	5	5
Cerastium holosteoides	6	6	6	6	5	6
Cerastium ligusticum	11	9	4	2	3	1
Cerastium semidecandrum	8	7	5	4	X	X
Minuartia hybrida subsp. *hybrida*	7	7	5	3	6	2
Minuartia mediterranea	11	9	4	2	3	2
Petrorhagia saxifraga	9	8	7	2	8	3
Polycarpon tetraphyllum subsp. *diphyllum*	7	7	5	4	5	6
Polycarpon tetraphyllum subsp. *tetraphyllum*	7	7	5	4	5	6
Saponaria officinalis	7	6	4	5	7	5
Silene colorata	11	9	3	1	X	1
Silene conica	9	7	5	2	5	2
Silene dioica	7	5	5	6	7	8
Silene italica	5	7	5	4	6	5
Silene latifolia subsp. *alba*	6	9	4	3	4	2
Silene vulgaris	8	X	X	4	7	2
Spergularia media	7	7	5	7	8	5
Stellaria media subsp. *media*	6	X	X	4	7	8

(continued)

Species	L	T	C	U	R	N
Stellaria pallida	8	8	5	3	5	4
Celastraceae						
Euonymus europaeus	6	5	5	5	8	5
Ceratophyllaceae						
Ceratophyllum demersum	6	7	X	12	8	8
Cistaceae						
Cistus creticus subsp. *eriocephalus*	11	9	4	2	3	2
Cistus salviifolius	11	9	4	2	2	2
Fumana procumbens	9	6	7	3	7	1
Helianthemum apenninum	9	7	3	2	7	2
Helianthemum jonium	11	10	4	2	7	1
Helianthemum nummularium	9	X	6	4	7	2
Commelinaceae						
Commelina communis	7	6	5	8	6	2
Convolvulaceae						
Calystegia sepium	8	6	5	6	7	9
Calystegia soldanella	11	8	4	1	X	1
Convolvulus arvensis	7	7	5	4	5	5
Cuscuta campestris	8	7	5	X	X	X
Cuscuta epithymum	8	X	5	X	X	X
Dichondra micrantha	5	8	5	6	3	2
Cornaceae						
Cornus mas	6	7	6	5	8	4
Cornus sanguinea subsp. *hungarica*	7	5	5	7	8	X
Crassulaceae						
Sedum acre	8	5	4	1	X	1
Sedum album	11	X	5	2	X	1
Cupressaceae						
Juniperus communis	8	0	0	4	0	4
Cyperaceae						
Bolboschoenus maritimus	8	X	4	10	8	5
Carex acutiformis	7	5	5	9	7	5
Carex caryophyllea	8	5	5	4	X	2
Carex distans	9	6	5	7	8	X
Carex divisa	8	8	2	3	5	3
Carex divulsa	7	6	5	4	5	5
Carex extensa	9	5	3	7	0	4
Carex flacca	7	5	5	6	8	X
Carex hirta	7	6	4	6	X	5
Carex liparocarpos	8	7	6	2	6	2
Carex otrubae	9	5	5	9	X	5
Carex panicea	8	4	4	7	X	3
Carex pendula	5	5	5	8	6	5

(continued)

Species	L	T	C	U	R	N
Carex punctata	7	6	3	10	4	3
Carex riparia	7	5	5	10	6	5
Carex rostrata	8	4	4	10	4	2
Carex viridula	8	X	5	8	X	2
Cladium mariscus	9	X	5	10	9	3
Cyperus capitatus	11	9	3	3	5	1
Cyperus eragrostis	8	9	5	10	6	6
Cyperus flavescens	6	6	5	9	5	5
Cyperus fuscus	6	6	5	9	5	5
Cyperus glomeratus	9	8	5	11	5	5
Cyperus laevigatus subsp. *distachyos*	9	11	5	9	6	5
Cyperus longus	8	7	5	11	5	5
Eleocharis palustris	8	6	5	10	3	3
Schoenoplectus lacustris	8	5	5	11	7	5
Schoenoplectus litoralis	9	8	4	10	6	6
Schoenoplectus mucronatus	8	8	4	10	7	8
Schoenoplectus pungens	8	6	5	10	7	7
Schoenoplectus tabernaemontani	8	7	6	10	8	X
Schoenoplectus triqueter	8	7	5	10	7	7
Schoenus nigricans	9	7	5	9	9	2
Scirpoides holoschoenus	9	9	4	8	7	4
Dennstaedtiaceae						
Pteridium aquilinum	6	5	4	6	3	3
Dioscoreaceae						
Dioscorea communis	5	7	5	5	8	6
Elaeagnaceae						
Elaeagnus angustifolia	9	7	5	3	X	2
Hippophae fluviatilis	9	6	7	7	X	2
Equisetaceae						
Equisetum arvense	6	X	X	6	X	3
Equisetum fluviatile	8	4	X	10	X	6
Equisetum palustre	7	X	5	7	X	3
Equisetum ramosissimum	7	7	6	3	7	1
Equisetum telmateia	5	7	4	8	8	5
Euphorbiaceae						
Euphorbia cyparissias	7	7	5	3	5	5
Euphorbia esula	7	6	5	7	5	5
Euphorbia helioscopia subsp. *helioscopia*	9	7	5	3	5	6
Euphorbia lucida	8	7	6	6	5	5
Euphorbia maculata	7	8	5	2	5	4
Euphorbia palustris	7	6	6	7	5	5
Euphorbia paralias	11	8	5	1	X	1
Euphorbia peplis	11	7	2	1	X	1

(continued)

Species	L	T	C	U	R	N
Euphorbia peplus	6	7	4	4	5	7
Euphorbia platyphyllos	6	7	5	5	5	6
Euphorbia prostrata	7	8	5	2	5	4
Mercurialis annua	7	7	5	4	7	8
Fabaceae						
Amorpha fruticosa	7	8	5	6	5	6
Astragalus glycyphyllos	7	6	6	4	7	4
Colutea arborescens	5	8	5	3	8	2
Dorycnium herbaceum	7	8	6	6	9	5
Dorycnium hirsutum	7	8	5	3	7	2
Dorycnium pentaphyllum	7	8	6	6	9	5
Genista tinctoria	5	6	5	5	3	3
Hippocrepis emerus	7	6	4	3	9	2
Lathyrus hirsutus	7	6	5	4	7	X
Lathyrus pratensis	7	5	X	6	7	6
Lathyrus sphaericus	11	9	5	2	5	2
Lotus angustissimus	11	8	5	7	7	4
Lotus corniculatus	7	X	5	4	7	2
Lotus maritimus	8	6	5	6	9	X
Lotus tenuis	9	7	5	6	7	7
Medicago lupulina	7	5	X	4	8	7
Medicago marina	12	8	5	1	X	1
Medicago minima	11	7	5	3	8	1
Medicago sativa subsp. *sativa*	8	5	7	3	9	3
Melilotus albus	9	6	6	3	7	3
Melilotus indicus	7	7	4	4	5	5
Melilotus neapolitanus	11	9	4	2	5	3
Melilotus officinalis	8	5	6	3	8	7
Ononis natrix subsp. *natrix*	8	8	5	3	8	0
Ononis spinosa	8	6	5	X	X	3
Robinia pseudoacacia	5	7	5	4	X	8
Trifolium arvense	8	5	5	2	2	1
Trifolium campestre	8	5	5	4	X	3
Trifolium fragiferum subsp. *fragiferum*	8	6	5	7	8	7
Trifolium nigrescens	8	6	5	5	5	6
Trifolium pratense	7	X	4	X	X	X
Trifolium repens	8	X	X	X	X	7
Trifolium scabrum subsp. *scabrum*	11	8	5	2	9	1
Trifolium striatum	8	8	5	3	2	0
Vicia cracca	7	X	X	5	X	X
Vicia sativa	5	5	6	X	X	X
Vicia villosa subsp. *ambigua*	11	9	4	3	7	2
Vicia villosa subsp. *varia*	7	6	5	4	4	5

(continued)

Species	L	T	C	U	R	N
Fagaceae						
Quercus ilex	2	9	4	3	X	X
Quercus pubescens subsp. *pubescens*	7	8	6	3	7	4
Quercus robur subsp. *robur*	7	6	6	6	5	6
Gentianaceae						
Blackstonia perfoliata	8	7	5	X	9	4
Centaurium erythraea	8	6	5	5	6	X
Centaurium pulchellum	9	6	7	7	9	3
Centaurium tenuiflorum	9	8	5	7	7	2
Schenkia spicata	11	9	5	3	7	3
Geraniaceae						
Erodium cicutarium	8	7	5	3	5	3
Geranium columbinum	7	9	6	2	5	2
Geranium dissectum	7	8	5	2	5	2
Geranium molle	7	6	5	3	5	4
Geranium purpureum	7	8	5	3	6	3
Geranium pusillum	7	7	5	4	5	6
Geranium rotundifolium	7	8	5	3	6	3
Haloragaceae						
Myriophyllum spicatum	5	X	X	12	8	5
Hydrocharitaceae						
Najas minor	6	8	4	12	8	4
Hypericaceae						
Hypericum perforatum subsp. *perforatum*	7	8	6	X	X	X
Iridaceae						
Iris pseudacorus	7	7	5	10	6	7
Juglandaceae						
Juglans regia	6	6	6	5	6	6
Juncaceae						
Juncus acutus	11	8	3	8	8	3
Juncus articulatus	8	7	4	8	6	5
Juncus bufonius	4	7	5	6	4	1
Juncus compressus	8	7	5	5	7	5
Juncus gerardii	8	6	4	5	7	5
Juncus inflexus	7	7	5	8	6	5
Juncus littoralis	11	8	3	8	8	3
Juncus maritimus	2	7	3	8	8	3
Juncus subnodulosus	8	6	4	9	6	5
Juncus tenageja	4	5	5	6	4	1
Luzula campestris	7	4	4	4	3	2
Juncaginaceae						
Triglochin maritimum	9	0	4	6	7	8
Lamiaceae						

(continued)

Species	L	T	C	U	R	N
Ajuga chamaepitys subsp. *chamaepitys*	7	8	5	4	9	2
Ajuga reptans	6	X	4	6	X	6
Ballota nigra subsp. *foetida*	8	6	5	5	X	8
Clinopodium acinos	11	X	5	2	7	1
Clinopodium nepeta	5	7	5	3	9	3
Clinopodium vulgare	7	5	4	4	7	3
Galeopsis ladanum	7	6	5	3	4	1
Glechoma hederacea	6	7	4	4	5	3
Lamium amplexicaule	7	7	5	4	5	7
Lamium purpureum	7	7	5	4	5	5
Lycopus europaeus	7	6	5	9	X	7
Marrubium vulgare	9	8	5	3	8	8
Mentha aquatica	7	5	5	9	7	4
Mentha suaveolens	7	8	5	8	7	6
Nepeta cataria	7	7	4	4	7	2
Origanum vulgare	7	6	5	3	X	3
Prunella laciniata	8	8	5	3	7	2
Prunella vulgaris subsp. *vulgaris*	7	6	4	6	4	X
Salvia pratensis	8	6	6	4	8	4
Salvia verbenaca	8	8	4	3	5	7
Stachys palustris	7	5	X	7	7	7
Stachys recta	7	6	4	3	8	2
Teucrium capitatum	11	8	4	2	X	1
Teucrium chamaedrys	7	6	5	2	8	1
Teucrium scordium subsp. *scordium*	7	7	5	8	8	2
Lauraceae						
Laurus nobilis	2	7	4	8	4	6
Lentibulariaceae						
Utricularia australis	9	6	4	12	5	4
Linaceae						
Linum bienne	7	7	5	3	7	2
Linum catharticum	7	X	5	X	X	1
Linum maritimum	11	8	4	2	7	2
Lythraceae						
Lythrum hyssopifolia	8	7	5	7	3	4
Lythrum salicaria	7	5	5	8	7	X
Malvaceae						
Abutilon theophrasti	8	9	6	7	5	4
Althaea officinalis	7	6	6	7	7	6
Kosteletzkya pentacarpos	11	8	6	7	7	7
Malva sylvestris	8	6	4	4	X	8
Moraceae						
Ficus carica	7	8	6	X	5	X

(continued)

Species	L	T	C	U	R	N
Morus alba	8	7	5	5	5	5
Nymphaeaceae						
Nymphaea alba	8	X	5	12	7	7
Oleaceae						
Fraxinus angustifolia subsp. *oxycarpa*	4	8	6	7	7	8
Fraxinus ornus	5	8	6	3	8	3
Ligustrum vulgare	7	6	4	X	8	X
Phillyrea angustifolia	11	10	4	1	X	2
Onagraceae						
Epilobium hirsutum	7	8	5	7	6	6
Epilobium parviflorum	7	6	5	8	6	6
Epilobium roseum subsp. *roseum*	7	5	5	8	8	8
Epilobium tetragonum	7	7	5	5	5	5
Oenothera stucchii	9	7	5	3	X	4
Orchidaceae						
Anacamptis coriophora	7	8	5	3	5	3
Anacamptis laxiflora	8	7	5	6	6	5
Anacamptis morio subsp. *morio*	7	5	4	4	7	3
Anacamptis palustris	8	7	5	7	8	6
Anacamptis pyramidalis	8	7	5	3	9	2
Cephalanthera damasonium	2	5	4	4	7	4
Cephalanthera longifolia	4	5	5	3	8	3
Cephalanthera rubra	3	5	5	4	8	3
Epipactis helleborine	3	5	5	5	7	5
Epipactis microphylla	7	7	4	3	6	2
Limodorum abortivum	X	7	5	4	8	0
Neotinea tridentata	8	6	5	3	6	3
Neottia nidus avis	2	5	5	5	7	5
Ophrys sphegodes subsp. *sphegodes*	8	8	5	4	9	3
Orobanchaceae						
Odontites vulgaris	6	X	5	5	X	X
Parentucellia latifolia	8	8	5	3	3	3
Parentucellia viscosa	8	8	3	3	3	3
Oxalidaceae						
Oxalis corniculata	7	7	0	4	X	6
Oxalis dillenii	7	7	5	5	5	7
Papaveraceae						
Chelidonium majus	6	6	X	5	X	8
Fumaria officinalis	7	7	5	4	5	6
Glaucium flavum	11	9	5	1	4	1
Papaver rhoeas	6	6	5	5	7	X
Phytolaccaceae						
Phytolacca americana	9	8	5	5	5	4

(continued)

Species	L	T	C	U	R	N
Pinaceae						
Pinus pinaster	11	8	4	2	4	3
Pinus pinea	11	8	5	2	4	3
Plantaginaceae						
Gratiola officinalis	7	7	5	9	5	5
Kickxia commutata	8	7	4	4	5	4
Linaria vulgaris subsp. *vulgaris*	8	5	5	3	7	3
Plantago afra	11	6	4	3	7	2
Plantago altissima	9	7	6	7	7	5
Plantago arenaria	8	7	6	3	7	4
Plantago coronopus	8	7	5	7	7	4
Plantago lanceolata	6	7	5	X	X	X
Plantago major	8	X	X	5	X	7
Plantago sempervirens	11	6	4	3	7	2
Veronica agrestis	5	4	4	6	7	7
Veronica anagallis-aquatica subsp. *anagallis-aquatica*	7	6	5	9	7	6
Veronica anagalloides	7	X	5	9	7	7
Veronica arvensis	5	5	5	5	6	X
Veronica catenata	7	6	4	9	7	6
Veronica hederifolia	6	6	5	5	3	7
Veronica officinalis	5	X	5	4	2	3
Veronica persica	8	7	5	5	5	6
Plumbaginaceae						
Limonium bellidifolium	11	10	4	1	9	1
Limonium narbonense	11	7	5	6	7	5
Limonium virgatum	9	9	3	1	9	1
Poaceae						
Aeluropus littoralis	11	10	4	4	8	1
Agrostis castellana	8	6	3	4	4	2
Agrostis stolonifera subsp. *stolonifera*	8	X	X	6	X	5
Aira elegantissima	8	9	5	2	3	1
Alopecurus myosuroides	6	6	5	6	7	7
Anthoxanthum odoratum	X	X	5	X	5	3
Arundo donax	8	9	5	5	5	6
Avena barbata	8	8	5	3	7	2
Brachypodium rupestre subsp. *caespitosum*	8	6	4	5	8	4
Brachypodium rupestre subsp. *rupestre*	8	6	4	5	8	4
Brachypodium sylvaticum	4	5	5	5	6	6
Bromus diandrus	8	8	5	3	5	4
Bromus erectus subsp. *erectus*	8	5	7	3	8	3
Bromus hordeaceus	7	6	5	X	X	X
Bromus madritensis	8	7	5	3	X	1
Bromus ramosus	6	5	5	6	8	6

(continued)

Species	L	T	C	U	R	N
Bromus squarrosus	8	9	5	2	4	2
Bromus sterilis	7	7	5	4	X	5
Bromus tectorum	8	6	7	3	8	4
Calamagrostis arundinacea	6	5	5	5	5	5
Calamagrostis epigejos	7	5	7	X	X	7
Calamagrostis pseudophragmites	7	6	5	8	8	6
Catapodium rigidum	8	8	5	2	5	4
Cenchrus spinifex	9	9	0	1	0	1
Corynephorus articulatus	8	9	4	2	3	1
Cynodon dactylon	8	8	5	4	X	4
Cynosurus cristatus	8	5	4	5	5	4
Dactylis glomerata	7	6	5	4	5	6
Danthonia decumbens	8	5	4	6	3	2
Dasypyrum villosum	8	10	5	2	4	2
Digitaria ciliaris	7	7	0	3	4	3
Digitaria sanguinalis	7	7	5	3	6	4
Echinochloa crusgalli	6	7	5	7	X	8
Eleusine indica	11	8	5	2	7	2
Elymus athericus	11	7	5	5	7	7
Elymus farctus	11	6	5	7	7	7
Elymus repens	7	X	7	5	X	8
Festuca pratensis	8	6	6	5	5	6
Festuca rubra	8	4	5	4	4	3
Glyceria maxima	9	5	5	10	8	7
Holcus lanatus	7	5	4	6	X	4
Hordeum marinum	11	9	3	3	6	3
Hordeum murinum subsp. *leporinum*	8	8	4	5	5	3
Hordeum secalinum	8	8	4	4	5	5
Lagurus ovatus	8	9	5	3	X	2
Lolium multiflorum	7	7	5	4	X	6
Lolium perenne	8	5	4	5	X	7
Molinia arundinacea	7	5	4	X	8	2
Panicum miliaceum	6	7	5	4	7	3
Parapholis incurva	11	7	4	5	7	2
Paspalum distichum	X	8	X	10	8	8
Phalaris arundinacea	7	X	X	8	7	7
Phleum arenarium subsp. *caesium*	8	6	4	2	7	3
Phragmites australis	7	5	X	10	7	5
Poa annua	7	X	5	6	X	8
Poa bulbosa	8	8	7	2	4	1
Poa pratensis	6	X	X	5	X	X
Poa sylvicola	3	5	5	5	4	6
Poa trivialis	6	X	5	7	X	7

(continued)

Species	L	T	C	U	R	N
Polypogon monspeliensis	8	8	5	9	8	6
Puccinellia distans	8	X	6	6	7	7
Puccinellia festuciformis	11	X	4	8	7	8
Rostraria cristata	7	5	5	6	8	2
Rostraria hispida	7	8	5	5	8	2
Saccharum ravennae subsp. *ravennae*	11	8	5	6	8	8
Scholochloa festucacea	9	8	5	6	8	6
Setaria pumila	7	7	5	4	5	6
Setaria verticillata	7	8	5	4	X	8
Sorghum halepense	8	8	X	7	8	8
Sporobolus anglicus	9	6	2	9	9	6
Sporobolus maritimus	9	6	2	9	9	6
Tragus racemosus	8	7	0	5	7	3
Trisetaria flavescens	7	X	5	X	X	5
Trisetaria michelii	11	9	4	2	5	1
Vulpia ciliata	8	9	5	2	4	2
Vulpia fasciculata	11	10	3	1	X	1
Vulpia myuros	8	9	5	2	6	2
Polygalaceae						
Polygala comosa	8	6	6	3	8	2
Polygala monspeliaca	8	8	4	5	7	1
Polygala nicaeensis subsp. *mediterranea*	8	6	5	3	7	2
Polygala vulgaris	7	4	5	5	3	2
Polygonaceae						
Fallopia convolvulus	8	7	4	4	5	3
Fallopia dumetorum	6	7	4	4	5	6
Persicaria lapathifolia	6	6	5	7	X	8
Persicaria maculosa	6	5	5	3	7	7
Persicaria minor	7	6	5	8	4	8
Polygonum arenastrum	7	8	5	3	6	1
Polygonum aviculare	7	7	5	3	6	1
Rumex crispus	7	5	5	6	X	5
Rumex cristatus subsp. *cristatus*	7	6	4	3	6	7
Rumex obtusifolius subsp. *obtusifolius*	7	5	6	3	X	9
Polypodiaceae						
Polypodium vulgare	5	X	4	X	2	X
Portulacaceae						
Portulaca oleracea	7	8	5	4	7	7
Potamogetonaceae						
Potamogeton crispus	6	5	5	12	7	6
Potamogeton nodosus	6	6	5	12	7	6
Potamogeton pusillus	6	5	5	12	7	8
Stuckenia pectinata	6	0	5	12	7	5

(continued)

Species	L	T	C	U	R	N
Zannichellia palustris subsp. *pedicellata*	6	0	2	12	7	6
Primulaceae						
Anagallis arvensis	6	6	5	5	X	6
Anagallis minima	7	7	5	7	4	2
Lysimachia vulgaris	7	X	7	9	X	X
Samolus valerandi	7	6	4	8	X	6
Ranunculaceae						
Clematis flammula	7	9	5	3	5	4
Clematis vitalba	7	7	4	5	7	7
Ficaria verna subsp. *fertilis*	4	5	5	6	7	7
Ranunculus acris	7	X	5	X	X	X
Ranunculus neapolitanus	8	6	5	3	7	3
Ranunculus peltatus subsp. *baudotii*	7	7	4	12	6	5
Ranunculus repens	6	X	X	7	X	7
Ranunculus sardous	8	7	5	8	X	7
Ranunculus sceleratus	9	X	X	9	7	9
Ranunculus serpens subsp. *polyanthemophyllus*	4	4	5	4	4	4
Ranunculus trichophyllus	7	X	X	12	X	X
Thalictrum lucidum	6	7	6	8	7	6
Rhamnaceae						
Frangula alnus subsp. *alnus*	6	5	4	7	5	5
Paliurus spina-christi	7	8	6	3	7	3
Rhamnus cathartica	7	5	5	4	8	X
Rosaceae						
Agrimonia eupatoria	7	6	5	4	8	4
Aphanes arvensis	6	5	5	6	4	5
Aphanes australis	7	7	4	5	4	4
Crataegus monogyna	6	7	5	4	6	3
Malus sylvestris	7	5	5	5	7	5
Potentilla erecta	6	X	5	X	X	2
Potentilla pusilla	8	6	4	2	8	0
Potentilla reptans	6	6	5	6	7	5
Potentilla tabernaemontani	7	5	5	2	7	2
Prunus spinosa subsp. *spinosa*	7	5	5	X	X	X
Pyracantha coccinea	5	8	4	3	5	3
Pyrus communis subsp. *pyraster*	6	5	5	6	7	7
Rosa canina	8	5	5	4	X	X
Rosa sempervirens	6	8	4	3	4	6
Rubus caesius	7	5	5	7	7	9
Rubus ulmifolius	5	8	5	4	5	8
Sanguisorba minor	7	6	5	3	8	2
Rubiaceae						
Asperula cynanchica	7	7	5	3	8	3

Species	L	T	C	U	R	N
Galium album	6	5	5	5	5	4
Galium aparine	6	X	5	4	5	5
Galium debile	7	5	5	8	5	3
Galium palustre	7	5	4	8	5	3
Galium verum	7	6	6	4	7	3
Rubia peregrina	5	9	4	4	5	3
Rubia tinctoria	7	7	4	5	5	5
Sherardia arvensis	8	6	5	5	8	5
Ruppiaceae						
Ruppia cirrhosa	7	8	3	12	7	7
Salicaceae						
Populus alba	5	8	7	5	8	6
Populus nigra	5	7	6	8	7	7
Populus × *canadensis*	5	7	X	7	X	6
Salix alba	5	6	6	7	8	7
Salviniaceae						
Azolla filiculoides	6	12	5	12	X	8
Salvinia natans	8	6	5	4	8	7
Santalaceae						
Osyris alba	7	8	5	3	4	2
Sapindaceae						
Acer campestre	5	7	4	5	7	6
Acer negundo	8	7	5	5	5	5
Acer platanoides	5	6	4	X	X	X
Saxifragaceae						
Saxifraga tridactylites	8	6	5	2	7	1
Scrophulariaceae						
Scrophularia canina	8	8	5	3	8	3
Verbascum blattaria	8	6	7	3	7	6
Verbascum densiflorum	8	6	5	4	8	5
Verbascum phlomoides	7	8	5	3	7	7
Verbascum sinuatum	9	8	5	3	7	7
Verbascum thapsus	8	X	4	4	7	7
Simaroubaceae						
Ailanthus altissima	6	7	5	5	5	5
Solanaceae						
Datura stramonium subsp. *stramonium*	9	8	5	3	5	7
Hyoscyamus albus	8	8	5	2	X	9
Solanum dulcamara	7	5	X	8	X	8
Solanum nigrum	7	6	5	3	5	7
Tamaricaceae						
Tamarix gallica	11	7	4	6	5	3
Thelypteridaceae						

(continued)

Species	L	T	C	U	R	N
Thelypteris palustris	5	X	X	8	5	6
Typhaceae						
Typha angustifolia	8	7	5	10	X	7
Typha latifolia	8	6	5	10	X	8
Ulmaceae						
Ulmus minor	5	7	5	X	8	X
Urticaceae						
Parietaria judaica	7	8	5	3	X	6
Urtica dioica subsp. *dioica*	X	X	X	6	X	8
Urtica membranacea	7	8	5	3	6	3
Urtica urens	7	6	X	5	7	8
Verbenaceae						
Verbena officinalis	9	5	5	4	X	6
Viburnaceae						
Sambucus ebulus	8	6	5	5	8	7
Sambucus nigra	7	5	4	5	X	9
Viburnum lantana	7	5	5	4	8	5
Viburnum opulus	6	5	5	7	7	6
Violaceae						
Viola alba subsp. *alba*	5	8	5	5	7	6
Viola alba subsp. *dehnhardtii*	5	8	5	5	7	6
Viola canina	7	5	5	4	3	2
Viola hirta	6	5	5	3	8	2
Viola odorata	5	6	5	5	X	8
Viola suavis subsp. *suavis*	5	8	6	5	4	4
Vitaceae						
Vitis × koberi	6	8	5	6	8	6
Vitis vinifera	6	8	5	6	8	6
Zygophyllaceae						
Tribulus terrestris	8	8	6	2	5	3

Appendix B: List of fungi recorded in the Bosco della Mesola Nature Reserve (from Bernicchia et Corbetta 1982; Comune di Mesola 1988)

Agaricaceae
Agaricus angustus Fr.
Agaricus benesii (Pilát) Pilát
Agaricus campestris L.
Agaricus depauperatus (F.H. Møller) Pilát
Agaricus fuscus-fibrillosus (F.H. Møller) Pilát
Agaricus impudicus (Rea) Pilát
Agaricus langei (F.H. Møller) F.H. Møller
Agaricus lanipes (F.H. Møller et Jul. Schäff.) Hlaváček
Agaricus phaeolepidotus F.H. Møller
Agaricus sylvaticus Schaeff.
Agaricus sylvicola (Vittad) Peck
Chamaemyces fracidus (Fr.) Donk
Cystolepiota pulverulenta (Huijsman) Vellinga
Cystolepiota seminuda (Lasch) Bon
Echinoderma asperum (Pers.) Bon
Lepiota boudieri Bres.
Lepiota brunneoincarnata Chodat et C. Martin
Lepiota forquignonii Quél
Lepiota helveola Bres.
Lepiota lilacea Bres.
Lepiota pseudofelina J.E. Lange
Macrolepiota fuliginosa (Barla) Bon
Tulostoma brumale Pers.
Tulostoma fimbriatum Fr.
Amanitaceae
Amanita boudieri Barla

(continued)

Amanita citrina Pers.
Amanita fulva Fr.
Amanita ovoidea (Bull.) Link
Amanita pantherina (D.C.) Krombh.
Amanita phalloides (Vaill. ex Fr.) Link
Amanita rubescens Pers.
Amanita strobiliformis (Paulet ex Vittard) Bertill.
Auriculariaceae
Auricularia auricula-judae (Bull.) Quél.
Auricularia mesenterica (Dicks.) Pers.
Exidia glandulosa (Bull.) Fr.
Bankeraceae
Phellodon niger (Fr.) P. Karst.
Boletaceae
Boletus reticulatus Schaeff.
Buchwaldoboletus hemichrysus (Berk. et M.A. Curtis) Pilát
Butyriboletus appendiculatus (Schaeff.) D. Arora et J.L. Frank
Leccinellum lepidum (H. Bouchet ex Essette) Bresinsky et Manfr. Binder
Leccinellum pseudoscabrum (Kallenb.) Mikšík,
Leccinum scabrum (Bull.) Gray
Rheubarbariboletus armeniacus (Quél.) Vizzini, Simonini et Gelardi
Botryobasidiaceae
Botryobasidium aureum Parmasto
Cantharellaceae
Cantharellus cibarius Fr.
Rickenella fibula (Bull.) Raithelh
Sistotrema brinkmannii (Bres.) J. Erikss.
Ceratiomyxaceae
Ceratiomyxa fruticulosa var. *fruticulosa* (O.F. Müll.) T. Macbr.
Coniophoraceae
Coniophora arida (Fr.) P. Karst.
Corticiaceae
Mutatoderma mutatum (Peck) C.E. Gómez
Vuilleminia comedens (Nees) Maire
Cortinariaceae
Cortinarius bulliardi (Pers.) Fr.
Cortinarius caligatus Malençon
Cortinarius salor Fr.
Cyphellaceae
Chondrostereum purpureum (Pers.) Pouzar
Dacrymycetaceae
Dacrymyces deliquescens (Bull.) Duby
Didymiaceae
Diderma spumarioides (Fr.). Fr.

(continued)

Dyatripaceae

Diatrypella quercina (Persoon ex Fries) Cooke

Entolomataceae

Entoloma araneosum (Quél) M.M. Moser

Entoloma incarnatofuscescens (Britzelm.) Noordel.

Fomitopsidaceae

Postia caesia (Schrad.) P. Karst.

Ganodermataceae

Ganoderma applanatum (Pers.) Pat.

Ganoderma lucidum (Curtis) P. Karst.

Geastraceae

Geastrum fornicatum (Huds.) Hook.

Geastrum pectinatum Pers.

Helvellaceae

Helvella acetabulum (L.) Quél.

Helvella crispa (Scop.) Fr.

Helvella fusca Gill.

Helvella monachella Scop. ex Fr.

Paxina queletii (Bres.) Stangl.

Hymenochaetaceae

Fomitiporia punctata (P. Karst.) Murrill

Fuscoporia ferruginosa (Schrad.) Murrill

Fuscoporia torulosa (Pers.) T. Wagner et M. Fisch.

Hymenochaete rubiginosa (Dicks.) Lév.

Hymenogastraceae

Gymnopilus junonius (Fr.) P.D. Orton

Hysteriaceae

Gloniopsis praelonga (Schwein.) Underw. et Earle

Hysterium angustatum Alb. et Schweinitz

Hydnodonthaceae

Brevicellicium olivascens (Bres.) K.H. Larss. et Hjortstam

Trechispora farinacea (Pers.) Liberta

Inocybaceae

Crepidotus variabilis (Pers.) Kumm.

Inocybe dunensis P.D. Orton

Inocybe pruinosa R. Heim

Inocybe tenebrosa Quél.

Mallocybe heimii (Bon) Matheny et Esteve-Rav.

Lachnocladiaceae

Vararia investiens (Schwein.) P. Karst.

Vararia ochroleuca (Bourdot et Galzin) Donk

Meruliaceae

Abortiporus biennis (Bull.) Singer

Hyphoderma roseocremeum (Bres.) Donk

(continued)

Hypochnicium punctulatum (Cooke) J. Erikss.
Junghuhnia nitida (Pers.) Ryvarden
Mycoacia fuscoatra (Fr.) Donk
Phlebia radiata Fr.
Phlebia rufa (Pers.) M.P. Christ.
Steccherinum fimbriatum (Pers.) J. Erikss.
Steccherinum ochraceum (Pers. Ex J.F. Gmelin) S.F. Gray
Steccherinum semisupiniforme (Murrill) Miettinen
Molliasiaceae
Tapesia fusca (Persoon ex Merat) Fuckel
Morchellaceae
Morchella crassipes (Vent.) Pers.
Morchella esculenta (L.) Pers.
Morchella umbrina Boud.
Verpa bohemica (Krbh.) Schröt
Verpa conica (O.F. Müll.) Sw.
Mycenaceae
Hemimycena cucullata (Pers.) Sing.
Hemimycena mauretanica (Maire) Singer
Mycena capyllaripes Peck.
Nitschkiaceae
Acanthonitschkea tristis (J. Kickxf.) Nannfeldt
Omphalotaceae
Gymnopus androsaceus (L.) Della Magg. et Trassin.
Gymnopus fusipes (Bull.) Gray
Paxillaceae
Paxillus rubicundulus P.D. Orton
Peniophoraceae
Peniophora cinerea (Pers.) Cooke
Peniophora incarnata (Pers.) P. Karst.
Peniophora quercina (Pers.) Cooke
Pezizaceae
Peziza phyllogena Cooke
Phallaceae
Clathrus ruber P. Micheli ex Pers.
Phanerochaetaceae
Byssomerulius corium (Pers.) Parmasto
Byssomerulius hirtellus (Burt) Parmasto
Efibula tuberculata (P. Karst.) Zmitr. et Spirin
Phanerochaete sordida (P. Karst.) J. Erikss. et Ryvarden
Phanerochaete velutina (DC.) P. Karst
Physalacriaceae
Cylindrobasidium evolvens (Fr.) Jülich
Desarmillaria tabescens (Scop.) R.A. Koch et Aime

(continued)

Hymenopellis radicata (Relhan) R.H. Petersen

Laccariopsis mediterranea (Pacioni et Lalli) Vizzini

Physaraceae

Fuligo septica (L.) F.H. Wigg.

Pleniophoraceae

Gloiothele lactescens (Berk.) Hjortstam

Pleurotaceae

Pleurotus cornucopiae (Paul.) Rolland

Pluteaceae

Pluteus cervinus (Schaeff) P. Kumm.

Volvariella caesiotincta P.D. Orton

Volvariella murinella (Quél.) M.M. Moser ex Dennis, P.D. Orton et Hora

Polyporaceae

Cerrena unicolor (Bull.) Murrill

Coriolopsis gallica (Fr.) Ryvarden

Cyanosporus subcaesius (A. David) B.K. Cui, L.L. Shen et Y.C. Dai

Daedaleopsis nitida (Durieu et Mont.) Zmitr. et Malysheva

Lentinus arcularius (Batsch) Zmitr.

Skeletocutis nivea (Jungh.) Jean Keller

Trametes trogii Berk.

Trametes versicolor (L.) Lloyd

Trichaptum fuscoviolaceum (Ehrenb.) Ryvarden

Tyromyces lacteus (Fr.) Murrill

Psathyrellaceae

Coprinellus micaceus (Bull.) Vilgalys, Hopple et Jacq. Johnson

Coprinellus xanthothrix (Romagn.) Vilgalys, Hopple et Jacq. Johnson

Coprinopsis picacea (Bull.) Redhead, Vilgalys et Moncalvo

Cystoagaricus hirtosquamulosus (Peck) Örstadius et E. Larss.

Parasola plicatilis (Curtis) Redhead, Vilgalys et Hopple

Radulomycetaceae

Radulomyces confluens (Fr.) M.P. Christ.

Radulomyces molaris (Chaillet ex Fr.) M.P. Christ.

Rickenellaceae

Peniophorella praetermissa (P. Karst.) K.H. Larss.

Russulaceae

Lactarius atlanticus Bon

Russula cyanoxantha (Schaeff.) Fr.

Russula delica Fr.

Russula emetica (Schaeff.) Pers.

Russula heterophylla (Fr.) Fr.

Russula vesca Fr.

Russula virescens (Schaeff.) Fr.

Schizophyllaceae

Schizophyllum amplum (Lév.) Nakasone

(continued)

Schizoporaceae

Fibrodontia gossypina Parmasto

Oxyporus latemarginatus (Durieu et Mont.) Donk

Oxyporus populinus (Schumach.) Donk

Schizopora paradoxa (Schrad.) Donk

Xylodon nespori (Bres.) Hjortstam et Ryvarden

Stephanosporaceae

Cristinia helvetica (Pers.) Parmasto

Stereaceae

Gloeocystidiellum porosum (Berk. et M.A. Curtis) Donk

Stereum gausapatum (Fr.) Fr.

Stereum hirsutum (Willd.) Pers.

Stereum sanguinolentum (Alb. et Schwein.) Fr.

Tapinellaceae

Tapinella atrotomentosa (Batsch) Šutara

Thelophoraceae

Thelephora ellisii (Sacc.) Zmitr., Shchepin, Volobuev et Myasnikov

Tomentella radiosa (P. Karst.) Rick

Tricholomataceae

Lepista nuda (Bull.) Cooke

Lepista personata (Fr.) Cooke

Omphalina pyxidata (Bull.) Quél.

Resupinatus applicatus (Batsch) Gray

Tubariaceae

Tubaria dispersa (Pers.) Singer

Tuberaceae

Rhizopogon magnatus (Picco) Corda

Tuber borchii Vittad.

Xenasmataceae

Xenasma pulverulentum (H.S. Jacks.) Donk

Xenasma tulasnelloideum (Höhn. et Litsch.) Donk

References

Agostini R (1965) Il Bosco della Mesola nei suoi aspetti selvicolturali. Notiziario della Società Italiana di Fitosociologia. Ann Bot (Roma) 28(2):6–11

Albanese R (1958) Il Gran Bosco della Mesola. Le Vie d'Italia 64(4):457–468

Alessandrini A, Balboni G, Piccoli F, Gerdol R, Pellizzari M, Brancaleoni L (2019) *Aphanes australis* Rydb., novità per la flora dell'Emilia-Romagna. Quad Mus civ St Nat Ferrara 7:37–40

Antolini G, Auteri L, Pavan V, Tomei F, Tomozeiu R, Marletto V (2016) A daily high-resolution gridded climatic data set for Emilia-Romagna, Italy, during 1961-2010. Int J Climatol 36:1970–1986

Argenti C, Masin R, Pellegrini B, Perazza G, Prosser F, Scortegagna S, Tasinazzo S (2019) Flora del Veneto dalle Dolomiti alla laguna veneziana. Cierre, Sommacampagna (VR)

Bartolucci F, Peruzzi L, Galasso G, Albano A, Alessandrini A, Ardenghi NMG, Astuti G, Bacchetta G, Ballelli S, Banfi E, Barberis G, Bernardo L, Bouvet D, Bovio M, Cecchi L, Di Pietro R, Domina G, Fascetti S, Fenu G, Festi F, Foggi B, Gallo L, Gottschlich G, Gubellini L, Iamonico D, Iberite M, Jiménez-Mejías P, Lattanzi E, Marchetti D, Martinetto E, Masin RR, Medagli P, Passalacqua NG, Peccenini S, Pennesi R, Pierini B, Poldini L, Prosser F, Raimondo FM, Roma-Marzio F, Rosati L, Santangelo A, Scoppola A, Scortegagna S, Selvaggi A, Selvi F, Soldano A, Stinca A, Wagensommer RP, Wilhalm T, Conti F (2018) An updated checklist of the vascular flora native to Italy. Plant Biosyst 152(2):179–303

Béguinot A (1910) Una escursione botanica nel littorale della Provincia di Ferrara. Boll Soc Bot Ital 9:125–136

Béguinot A (1916) I distretti floristici della regione littoranea dei territori circumadriatici. Schizzo fitogeografico. Rivista Geografica Italiana 23(2–4):1–44

Bernicchia A (1983) *Meruliopsis hirtellus* (Burt) Ginns, specie rara in Europa e nuova per l'Italia. Micologia Italiana 12(1):29–32

Bernicchia A (1986) *Phellinus punctatus* al Bosco della Mesola. Monti e Boschi 37(2):45–46

Bernicchia A, Corbetta F (1982) *Aphyllophorales* del Bosco della Mesola (Ferrara). Giorn Bot It 116:255–268

Bernicchia A, Intini M (1988) Deperimento del Bosco della Mesola associato alla presenza di agenti di carie e marciumi radicali. Monti e Boschi 39(5):53–57

Bernicchia A, Corbetta F, Padovan F (1987) Prime osservazioni micofitosociologiche al Bosco della Mesola (Ferrara). In: Pacioni G (ed) Studies on fungal communities, L'Aquila, pp 119–140

Bini C, Buffa G, Gamper U, Sburlino G, Zuccarello V (2002) Alcune considerazioni sui rapporti fra fitosociologia e pedologia. Fitosociologia 39(1):71–80

© The Author(s), under exclusive license to Springer Nature Switzerland AG 2021   103
A. Alessandrini et al., *The Vascular Flora of the Bosco della Mesola Nature Reserve (Northern Italy)*, Geobotany Studies, https://doi.org/10.1007/978-3-030-63412-4

Blasi C, Di Pietro R, Filesi P (2004) Syntaxonomical revision of *Quercetalia pubescenti-petraeae* in the Italian Peninsula. Fitosociologia 41(1):87–164

Bonani S (1989) Bosco della Mesola-Natura e storia. Oasis 5(9):70–85

Bondesan M, Castiglioni GB, Elmi C, Gabbianelli G, Marocco R, Pirazzoli PA, Tomasin A (1995) Coastal areas at risk from storm surges and sea-level rise in Northeastern Italy. J Coastal Res 11 (4):1354–1379

Brancaleoni L, Gerdol R, Abeli T, Corli A, Rossi G, Orsenigo S (2018) Nursery pre-treatment positively affects reintroduced plant performance via plant pre-conditioning, but not via maternal effects. Aquatic Conserv Mar Freshw Ecosyst 28(3):641–650

Buffa G, Ghirelli L, Lorenzoni GG (1994) Significato dei relitti vegetazionali a *Quercus ilex* L. nella valutazione della mediterraneità. Atti Conv Gr Bioritmi Vegetali e Fenologia S.B.I., Lecce, Orantes, pp 191–197

Busnardo G (2000) Segnalazioni floristiche per il Veneto centro-orientale. Ann Mus Civ Rovereto, Sez Arch St Sc Nat 15:83–105

Caramalli C (1987a) Le riserve naturali del Parco del Delta. Atti del Convegno Verde Ferrara. Incontro sul verde 'vivo'. Ferrara 7-8 Marzo 1985: 67–76. Istituto Gramsci Ferrara, Lega Ambiente. Corbo, Ferrara

Caramalli C (1987b) Il Gran Bosco. La riserva naturale Bosco della Mesola. In: Mesola: il Castello/il Gran Bosco. SIACA, Cento (FE)

Carullo F (1953) Lo storico bosco della Mesola. Monti e Boschi 4(11–12):487–504

Cencini C (1979) I boschi della fascia costiera emiliano-romagnola. In: Menegatti B (ed) Ricerche geografiche sulle pianure orientali dell'Emilia-Romagna. Patron, Bologna, pp 55–110

Comune di Mesola (1988) Funghi del Bosco della Mesola. Grafiche Zanini, Anzola dell'Emilia (BO)

Corbau C, Simeoni U, Zoccarato C, Mantovani G, Teatini P (2019) Coupling land use evolution and subsidence in the Po Delta, Italy: revising the past occurrence and prospecting the future management challenges. Sci Total Environ 654:196–1208

Corbetta F (1973) Che vogliamo fare del Bosco della Mesola? Ecologia 8:9–17

Corbetta F, Pettener D (1976) Lineamenti vegetazionali del Bosco della Mesola. Giorn Bot It 110 (6):448–449

Cori G, Raminelli G (1982) Mesola, Massenzatica, Monticelli - Pagine di storia del Mesolano. Arti Grafiche Masini, Serravalle (FE)

Cortesi P (2005) Il Boscone della Mesola. Minerva, Argelato (BO)

Debolini P, Ricceri C (1976) Una novità per la flora italiana: *Euphorbia lucida* Waldst. et Kit. Webbia 30:285–293

Domina G, Galasso G, Bartolucci F, Guarino R (2018) Ellenberg indicator values for the vascular flora alien to Italy. Flora Mediterr 28:53–61

Ellenberg H, Mueller-Dombois D (1965/1966) A key to Raunkiaer life forms with revised subdivisions. Ber Geobot Inst ETH Stiftung Rübel 37:56–73

Fattorini S (2017) The Watson–Forbes biogeographical controversy untangled 170 years later. J Hist Biol 50:473–496

Ferrari C (1994) I querceti misti della Pianura Padana sudorientale. Storia urbana 69:9–22

Galasso G, Conti F, Peruzzi L, Ardenghi NMG, Banfi E, CelestiGrapow L, Albano A, Alessandrini A, Bacchetta G, Ballelli S, Bandini Mazzanti M, Barberis G, Bernardo L, Blasi C, Bouvet D, Bovio M, Cecchi L, Del Guacchio E, Domina G, Fascetti S, Gallo L, Gubellini L, Guiggi A, Iamonico D, Iberite M, Jiménez-Mejías P, Lattanzi E, Marchetti D, Martinetto E, Masin RR, Medagli P, Passalacqua NG, Peccenini S, Pennesi R, Pierini B, Podda L, Poldini L, Prosser F, Raimondo FM, Roma-Marzio F, Rosati L, Santangelo A, Scoppola A, Scortegagna S, Selvaggi A, Selvi F, Soldano A, Stinca A, Wagensommer RP, Wilhalm T, Bartolucci F (2018) An updated checklist of the vascular flora alien to Italy. Plant Biosyst 152(3):556–592

Gamper U, Filesi L, Buffa G, Sburlino G (2008) Diversità fitocenotica delle dune costiere nord-adriatiche 1- le comunità fanerofitiche. Fitosociologia 45(1):3–21

Gerdol R, Ferrari C, Piccoli F (1985) Correlation between soil characters and forest types: a study in multiple discriminant analysis. Vegetatio 60:49–56

Gerdol R, Brancaleoni L, Lastrucci L, Nobili G, Pellizzari M, Ravaglioli M, Viciani D (2018) Wetland plant diversity in a coastal nature reserve in Italy: relationships with salinization and eutrophication and implications for nature conservation. Estuaries Coasts 41(7):2079–2091

Ginns J, Bernicchia A (1984) *Flaviporus semisupiniformis* (Polyporaceae) in Italy. Mycol Helv 1 (3):185–188

Gottfried M, Pauli H, Futschik A, Akhalkatsi M, Barančok P, Benito Alonso JL, Coldea G, Dick J, Erschbamer B, Fernández Calzado MR, Kazakis G, Krajči J, Larsson P, Mallaun M, Michelsen O, Moiseev D, Moiseev P, Molau U, Merzouki A, Nagy L, Nakhutsrishvili G, Pedersen B, Pelino G, Puscas M, Rossi G, Stanisci A, Theurillat JP, Tomaselli M, Villar L, Vittoz P, Vogiatzakis I, Grabherr G (2012) Continent-wide response of mountain vegetation to climate change. Nat Clim Change 2(2):111–115

IDROSER (1985) Analisi dell'ecosistema Bosco della Mesola - Valle della Falce e definizione di un sistema di controllo per una gestione ottimale, Relazione Generale. IDROSER, Bologna

Jedlowski E (1960) Profilo delle foreste demaniali di Bologna e di Ferrara - Il Gran Bosco della Mesola. Monti e Boschi 2(11):540–548

Longhi G (1967) Piano economico della foresta demaniale "Bosco della Mesola" per il decennio 1967-76. Technical Report

Longhi G (1968) La foresta demaniale di Bosco Mesola. Natura e Montagna s.3 8(4):41–46

Masin R, Scortegagna S (2012) Flora alloctona del Veneto centro-meridionale (province di Padova, Rovigo, Venezia e Vicenza – Veneto – NE Italia). Natura Vicentina 15:5–54

Mazzotti S, Pellizzari M, Piccoli F, Volponi S (2003a) I boschi planiziali e ripariali. In: Mazzotti S (ed) Biodiversità in Emilia-Romagna. Dalla biodiversità regionale a quella globale. Museo Civico di Storia Naturale di Ferrara, Regione Emilia-Romagna, pp 65–70

Mazzotti S, Pellizzari M, Piccoli F (2003b) I boschi costieri. In: Mazzotti S (ed) Biodiversità in Emilia-Romagna. Dalla biodiversità regionale a quella globale. Museo Civico di Storia Naturale di Ferrara, Regione Emilia-Romagna, pp 85–90

Mazzotti S, Montanari F, Pellizzari M, Cavalieri D'Oro A (2007) L'area di studio e i siti di campionamento. Quad Staz Ecol civ Mus St nat Ferrara 17:19–33

Merloni N, Piccoli F (2001) La vegetazione del complesso Punte Alberete e Valle Mandriole (Parco Regionale del Delta del Po - Italia). Braun-Blanquetia 29:1–17

Minerbi B (ed) (1984) Riserva Naturale Gran bosco della Mesola. Piano di Gestione Naturalistica per il decennio 1980-1989. Ministero Agricoltura e Foreste, Roma

Minerbi B (1985) Il Gran Bosco della Mesola. Economia Montana 17(5):16–22

Minerbi B, Leporati L, Corbetta F (1975) Il Boscone della Mesola. In: Le foreste in Emilia-Romagna. Regione Emilia-Romagna, Calderini, Bologna

Naccarato G (ed) (2004) Piano di gestione naturalistica della Riserva Naturale dello Stato "Bosco della Mesola". Corpo Forestale dello Stato - Ufficio Amministrazione Gestione ex A.S.F.D. di Punta Marina – Parco del Delta del Po Emilia – Romagna. Technical Report

Orsenigo S, Fenu G, Gargano D, Montagnani C, Abeli T, Alessandrini A, Bacchetta G, Bartolucci F, Carta A, Castello M, Cogoni D, Conti F, Domina G, Foggi B, Gennai M, Gigante D, Iberite M, Peruzzi L, Pinna MS, Prosser F, Santangelo A, Selvaggi A, Stinca A, Villani M, Wagensommer RP, Tartaglini N, Duprè E, Blasi C, Rossi G (2020) Red list of threatened vascular plants in Italy. Plant Biosyst. https://doi.org/10.1080/11263504.2020. 1739165

Pagnoni GA (1998) Mesola - il territorio e il Boscone. Una guida tra storia e natura. Corbo, Ferrara

Pagnoni GA (1999) Mesola e il suo Boscone. Laguna 3:16–23

Patrone G (1948) Piano di assestamento del Gran Bosco della Mesola per il trentennio 1948–1977. Unpublished

Patrone G (1951) Piano di assestamento del Gran Bosco della Mesola. In: Esempi di piani di assestamento forestale, Firenze, pp 5–11

Pedrotti F, Gafta D (1996) Ecologia delle foreste ripariali e paludose dell'Italia. L'Uomo e l'Ambiente 23:1–165

Pellizzari M, Brancaleoni L (2016) Note sulla flora ferrarese: il genere *Viola* L. (Violaceae). Quad Mus civ St Nat Ferrara 4:43–52

Pellizzari M, Pagnoni GA (1998) Flora e vegetazione nella Stazione Volano-Mesola-Goro. In: Il Piano Territoriale del Parco: analisi faunistica e floristico-vegetazionale. Atti Giornata di studio Qualità Ambientale nel Parco del Delta del Po, Palazzo Bellini, Comacchio (FE) Dicembre 1997: 125–132. Consorzio Parco Regionale del Delta del Po. CDS, Ferrara

Pellizzari M, Piccoli F (2001) La vegetazione dei corpi idrici del Bosco della Mesola (Delta del Po). Quad Staz Ecol civ Mus St nat Ferrara 13:7–24

Piccoli F (1964) Contributo allo studio floristico del Bosco della Mesola. Unpublished degree thesis, University of Ferrara

Piccoli F (1987) Tipologia della vegetazione boschiva nel Ferrarese. In: Atti del Convegno Verde Ferrara, Incontro sul verde 'vivo'. Ferrara 7-8 Marzo 1985:102–109, Istituto Gramsci Ferrara, Lega Ambiente. Corbo, Ferrara

Piccoli F (1995) Elementi per una carta della vegetazione del Parco Regionale del Delta del Po. (Regione Emilia-Romagna). Fitosociologia 30:213–219

Piccoli F (2005) Il Gran Bosco della Mesola: vestigio di antiche selve litoranee. In: Il Boscone della Mesola. Minerva, Argelato (BO), pp 21–24

Piccoli F, Gerdol R (1980) Typology and dynamics of a wood in the Po Plane (N-Italy): the Bosco della Mesola. Coll Phytosoc 9:161–170

Piccoli F, Gerdol R (1990) La vegetazione attuale. In: Il Parco del delta del Po, studi ed immagini. 1. L'ambiente come risorsa. Spazio Libri, Ferrara, pp 87–106

Piccoli F, Pellizzari M (2003) Note ecologiche sulle comunità pleustofitiche a *Lemna minuta* H., B. et K. nel Parco Regionale del Delta del Po. Atti del Convegno 'Botanica delle zone umide' Vercelli 10-11 novembre 2000. Museo Regionale di Scienze Naturali Torino, pp 221–230

Piccoli F, Gerdol R, Ferrari C (1983) Carta della vegetazione del Bosco della Mesola (Ferrara). Atti dell'Istituto Botanico e Laboratorio Crittogamico dell'Università di Pavia 2:3–23

Piccoli F, Corticelli S, Dell'Aquila L, Merloni N, Pellizzari M (1996) Vegetation map of the Regional Park of the Po Delta (Emilia-Romagna Region). Allionia 34:325–331

Piccoli F, Dell'Aquila L, Pellizzari M (1999) Carta della vegetazione del Parco Regionale del Delta del Po. Stazione Volano - Mesola - Goro. Scala 1:35.000. Regione Emilia - Romagna, Servizio Cartografico e Geologico

Piccoli F, Pellizzari M, Alessandrini A (2014) Flora del FerrareseIstituto per i Beni Artistici, Culturali e Naturali della Regione Emilia-Romagna. Longo, Ravenna

Pignatti S (2005) Valori di bioindicazione delle piante vascolari della flora vascolare d'Italia. Braun-Blanquetia 39:3–97

Raminelli G, Mantovani E, Rossi M, Bui D (1993) Mesola - la Storia, il Territorio, l'Ambiente. Club Alpino Italiano - Sezione di Ferrara, Cartografica Artigiana, Ferrara

Revedin P (1909) Contributo alla flora vascolare della provincia di Ferrara. Nuovo Giorn Bot It 16:269–334

Rossi M (1994) La Tenuta di Mesola: il recinto della bonifica. Anecdota 5(1):23–41

Simeoni U, Corbau C (2009) A review of the Delta Po evolution (Italy) related to climatic changes and human impacts. Geomorphology 107:64–71

Soil Survey Staff (1960) Soil classification, a comprehensive system. Soil Conservation Service, Washington

Stampi P (1966a) Il Gran Bosco della Mesola (Ferrara): notizie storiche, floristiche e geobotaniche. Ann Bot (Roma) 28(3):599–612

Stampi P (1966b) Le antiche selve che vegetano sul litorale ferrarese (Nota preventiva). Giorn Bot It 73:337–339

Stanisci A, Presti G, Blasi C (1998) I boschi igrofili del Parco Nazionale del Circeo. Ecol Mediterr 24:73–88

Stefani M, Vincenzi S (2005) The interplay of eustasy, climate and human activity in the late Quaternary depositional evolution and sedimentary architecture of the Po Delta system. Mar Geol 222–223(5–6):19–48

Taffetani F, Biondi E (1995) Boschi a *Quercus cerris* L. e *Carpinus orientalis* Miller nel versante Adriatico italiano. Ann Bot (Roma) 51 Suppl 10(2):229–240

Tiner RW (1999) Wetland indicators. A guide to wetland identification, delineation, classification, and mapping. Lewis, Boca Raton

Zangheri P (1971) La vegetazione del litorale emiliano – romagnolo nel passato e nel presente. Atti Convegno Italia Nostra. I Beni Naturali del litorale emiliano-romagnolo: problemi e prospettive. Ferrara, pp 24-36

Zemello M (1991) Il Bosco della Mesola ed il territorio circostante. Analisi storico-cartografica degli aspetti vegetazionali e fisiografici. Unpublished degree thesis, University of Padova